DATE DUE

THE BODY REVEALS

PHOTOGRAPHY—FRANK DESCISCIOLO

ILLUSTRATIONS—MICK BRADY and LYNDA BRAUN

FOREWORD—JOHN C. LILLY, M.D.

INTRODUCTION—DANIEL GOLEMAN, Ph.D.

THE BODY REVEALS

HOW TO READ YOUR OWN BODY

RON KURTZ AND
HECTOR PRESTERA, M.D.

A QUICKSILVER BOOK

1817

HARPER & ROW, PUBLISHERS, San Francisco
Cambridge, Hagerstown, New York, Philadelphia
London, Mexico City, São Paulo, Sydney

"The Silken Tent" from *The Poetry of Robert Frost* edited by Edward Connery Lathem. Copyright 1942 by Robert Frost. Copyright © 1969 by Holt, Rinehart and Winston, Inc. Copyright © 1970 by Lesley Frost Ballantine. Reprinted by permission of Holt, Rinehart and Winston, Publishers.

Library of Congress Cataloging in Publication Data

Kurtz, Ron.
  THE BODY REVEALS.

  "A Quicksilver book."
  Reprint. Originally published: New York : Harper & Row, c1976.
  Bibliography: p.
  1. Mind and body therapies.   2. Nonverbal communication (Psychology)
I. Prestera, Hector.   II. Title.
RC489.M53K87   1984   155.2   83-13016
ISBN 0-06-250488-6

89 90 91 92 10 9 8 7

This book is dedicated to our parents,
to Ida P. Rolf and Wilhelm Reich,
our teachers
in the flesh and in the spirit,
and to the
Great Teacher
Whose patience is indeed
Infinite.

CONTENTS

FOREWORD

For an understanding of the silent communication between bodies, I recommend this book. Without benefit of voicing or whispering, without the sounds of the larynx, pharynx, tongue, lips, teeth excited by expired gases traveling above critical velocities, the body transmits messages to those trained to receive them. Beyond/under phons/words/sentences/utterances, each of our bodies says who/what/where/how each of the residents in the body is/was/will be.

To change these body-messages to a closer correspondence with that which we can become, we change the resident-in-the-body and the body. The messages then change to correspond to the new form/substance of the resident. All such changes and their corresponding messages are slow. They take time, awareness and intent.

Two of the men/women deciphering this "silent language" are Hector Prestera and Ron Kurtz. Hector is a unique M.D. He and I shared life together for some time at Esalen Institute, Big Sur, California. We mutually learned and taught ourselves and others. We shared exciting changes in ourselves, our bodies, our minds, our loves, our antipathies, our attachments, our aversions.

Over the years since then we have stayed in contact, apart yet still together in spirit. His is the explorer-healer domain; mine that of the explorer-teacher. Yet we overlap enough in our domains to remain close companions in our journey through the samsaric existence.

Hector has been most fortunate in finding Ron Kurtz as a traveling companion in mapping new areas of healing. Ron is exploring healing by seeing and utilizing the unity of body, mind and spirit. My meeting with him was brief though in it I felt his warmth and openness of mind.

Both Hector and Ron have traveled far beyond this book. The primitive mechanical nature of publishing books always leaves fast travelers way ahead of where they were when they wrote them. Currently they continue their explorations of healing which are not yet part of the medical curriculum, though through efforts such as theirs, new areas will open in the training of physicians.

JOHN C. LILLY, M.D.

PREFACE TO THE SECOND EDITION

For this edition we have changed practically nothing. We have added to and updated the bibliography, and we have changed the cover. That's about it. We feel strongly that the book and its message are still timely and complete.

In the nine years since the first edition, we have both been working and growing. Ron has been giving workshops in the U.S. and Europe, teaching and writing about Body-centered Psychotherapy, and he started the Hakomi Institute in Boulder to teach and promote this work. Hector has continued to develop his awareness of the usefulness of Oriental medicine (acupuncture in particular) in terms of integrating the mind and the body. He has been working on a self-programming system that he has presented in several seminars under the heading of "Reprinting." In addition to seminar presentation, he is looking forward to the publication of this material as he deepens his experience with this process.

It is with great pleasure that we see this book go into a second edition. Thanks again to all the folks who helped.

HECTOR PRESTERA, M.D.
Monterey, California

RON KURTZ
Boulder, Colorado

PREFACE

The focus of therapeutic practice has been changing dramatically, away from the purely verbal approaches of the previous decades towards *action* and *doing*. No longer will patient and therapist just talk. Since the new approaches are primarily experiential and body-oriented, the future more and more will focus on the body.

We have written a book that we hope will be a simple, useful introduction to what body structure, posture, and physiognomy reveal about people. For those who can see and understand, the body speaks clearly, revealing character and a person's way of being in the world. It reveals past trauma and present personality, feelings expressed and feelings unexpressed. The trained eye sees this and much more.

Though much knowledge is available about what bodies reveal, it is not widely known. And while some use this knowledge intensively, it is not widely used. So it is our intention to present a basic outline of that knowledge, without subtlety or lengthy exposition. We want it to be useful to both lay people and those who heal. We hope mainly to make this point: the body reveals the person; it *is* the person. And we wish to bring that awareness to a wider audience.

In the chapters of this book we first discuss what the body reveals, what can be learned about a person by looking at their posture and body structure. Then we cover such basic concepts as energy, gravity, grounding, etc., the theoretical and philosophical underpinnings of the body-oriented approach. In the chapter "Body Parts" we cover the body section by section, giving detailed information on the relationship between the physical aspects of the body and the emotions and attitudes to which they correspond. In "Five People," we look at five individuals and give short descriptions of what one can tell from their bodies. This demonstrates the typical approach to reading whole, live people. Then we take a look at our own bodies in detail. We read our bodies and discuss them in a long transcript with comments.

Next, we provide a chapter which the reader can use to discover what his own body reveals, with instructions, illustrations and tables for that purpose. Finally, in "Last Thoughts," we discuss some the main approaches to body-oriented therapies. We get a nice assist there from such practitioners as Ilana Rubenfeld, Betty Fuller, and Richard Wheeler who discuss their work.

Every author faces and, certainly in this case, wholeheartedly enjoys the task of giving credit to those who have furthered his understanding and growth. They are surely too many to list separately. We hope to partially fulfill this pleasant obligation by mentioning a few.

Because of their great contribution to the understanding of the part played by the body in emotional illness and well-being, we are indebted to the pioneering work of Wilhelm Reich and Ida Rolf. Fritz Perls, Alexander Lowen, John Pierrakos, Moshe Feldenkrais, Judith Aston, Professor Jack Worsley, F. Matthias Alexander, Charlotte Selver, Will Schutz, and Randolph Stone also added immeasurably to our knowledge of the body/mind. For their personal help, we'd like to thank John Lilly, John Pierrakos, Pat Terry, Rosemary Feitis, Dick Price, John Heider, the late George Simon, Ron Robbins, Ken Lux, Philo Farnsworth III, Bill Solomon, Sam Pasiencier, who contributed to our conversations and offered stimulating insights; Ilana Rubenfeld, Betty Fuller, and Rick Wheeler, who contributed to the chapter on body-oriented therapies; our publisher and good friend, Bob Silverstein, who nursed and guided two fledglings through their first book; Frank Descisciolo, who did the photographs and developed the process that made the silhouettes; Mick Brady and Lynda Braun, who did the other illustrations; and all the gang around both our houses, Sharon, Tory, Derek, Susan, Rita, Lynda, and Jennie, for their concern, support, patience, and wisdom.

Finally, we'd like to express our deep appreciation to our patients, generous and courageous people, who offered themselves as photographic subjects for this book.

The information in this book can be a tool to promote a deeper understanding of people and a closer contact between them. We truly hope it will be.

HECTOR PRESTERA, M.D.
RON KURTZ

INTRODUCTION

One sunny day not long ago, I had the pleasure of sitting with Ron Kurtz in an outdoor cafe on a busy street in Boulder, Colorado. The sun had brought out a crowd, and the human species was on parade. From our viewing stand Ron guided me through the inner meaning of that parade, reading bodies as they passed. A master at the game, he showed me that day what the body does, indeed, reveal.

What the body reveals—at least when read by an expert like Ron—are the bends and twists of personal history: the secrets, traumas, and triumphs of days past. They are embodied in ligament and muscle; stamped into posture. The body is the personality made manifest and can be analyzed just as surely as can the psyche by those who know how.

Hector Prestera and Ron Kurtz are pioneers in reading what the body has to say about the person. In this now-classic volume, they give us a manual for deciphering the messages encoded in the body. As more and more therapists come to understand that the body does not lie, while words often do, reading the body is becoming the new royal road to the unconscious—and to all the other beliefs, feelings, and memories that shape and organize behavior. This book is an excellent map for following that road.

For those of us who are new to reading bodies, *The Body Reveals* is both delight and enlightenment. It gives the reader a guided tour of the body: part by part and in terms of attitudes and feelings that correspond to body structure. We then meet five people and see the stories their bodies tell about them. Finally, we learn how to apply the art of body reading to ourselves. The silhouettes of real bodies that illustrate each point make this guidebook eminently practical.

When *The Body Reveals* was first written, there was nothing like it. It became the Bible for the growing numbers of therapists whose invitation to clients was not, "say whatever comes into your mind," but rather, "stand up and let's have a look at you." There is, for all practical purposes, still no guidebook to the body quite like *The Body Reveals*. It stands as the primer, the essential introduction, to body-based therapies.

Once you acknowledge the intimate link between mind and body, the rationale for body therapies is clear. Nothing the muscles do is done without direction from the central nervous system; even the long-term habits of movement and holding that sculpt the body are put into place by the brain. The body is mind made manifest. To work on the body is to probe the mind.

The body, in turn, can become part of the elaborate system of defenses that the psyche raises to protect itself from the raw stuff of life. Character is armor and the body is the first line of defense. In the ways habit limits movement, curtails flexibility, and otherwise blocks and truncates our experience, the body armors us against the full experience of life. Body therapies are a direct way to break through that armor, to open fully to life.

Prestera and Kurtz take these assumptions as a starting point. *The Body Reveals* details the ties between mind and body in a way that makes those ties of use to any of us who want to unite mind and body into a cohesive organ, a whole.

—DANIEL GOLEMAN, PH.D.
Senior Editor, *Psychology Today*

THE BODY REVEALS

1.

WHAT THE
BODY REVEALS

The body never lies. Its tone, color, posture, proportions, movements, tensions, and vitality express the person within. These signs are a clear language to those who have learned to read them. The body says things about one's emotional history and deepest feelings, one's character and personality. The reckless staggering of a drunk and the light, graceful walk of a dancer speak as much to their movement through life as to their progress through space.

It is often easy to recognize a person by his walk, even at a distance. In doing so, we use the same cues that tell us of his life-style. A drooping head, slumped shoulders, a caved-in chest, and a slow, burdened gait reflect feelings of weakness and defeat, while a head carried erect, shoulders straight and loose, a chest breathing fully and easily, and a light gait tell of energy and confident promise. Such physical patterns become fixed by time, effecting growth and body structure, and characterizing not just the moment, but the person. Rather than simply a present disappointment, the crushed posture of hopelessness could be pointing to a lifetime of endless frustration and bitter failure.

Fixed muscular patterns in the body are central to a person's way of being in the world. They form in response to family and early environment. A child can be treated in many different ways by parents. If the general attitude toward him

or her is one which produces painful feelings, such as abandonment: "Go away! We don't want you!"; or disparagement: "You never do anything right," then the child develops a characteristic response and mood. It might be a mood of bitterness and a response of "I don't need anybody," or "I'll show them." Whatever the feeling, it is also expressed physically, and becomes a way of holding oneself, a fixed muscular pattern and a set attitude toward life. These are very likely to persist if nothing is done to change them.

These attitudes and fixed muscular patterns reflect, enhance, and sustain one another. It is as if the body sees what the mind believes and the heart feels, and adjusts itself accordingly. This gives rise to a way of holding oneself, as pride can swell the chest or fear contract the shoulders. The muscular pattern in turn sustains the attitude, as for example slouching forward, which makes every action more difficult and so makes life itself seem burdensome.

The mother's feelings for her child, and her responses to his or her physical and emotional needs, are the most important determinants of fixed patterns. A mother who responds to these needs with love and understanding helps the child experience security, satisfaction, and pleasure. These good feelings promote healthy, unimpeded growth and are a form of nourishment as important as food. Lack of this nourishment is crippling. As much as anything else, it is this crippling which the body reveals.

Ideally, the body is capable of allowing the free flowing of any feeling. It is efficient and graceful in its movements, aware and responsive to real needs. Such a body has bright eyes, breathes freely, is smooth skinned, and has an elastic muscle tone. It is well proportioned, and the various segments coordinate with each other. The neck is pliable and the head moves easily. The pelvis swings freely. The entire body is lined up efficiently with respect to gravity; that is, in a standing position, there is no struggle with gravity's downward pull. Pleasure and well-being are the characteristic feelings. A person with such a body is emotionally flexible and his or her feelings are spontaneous.

The various fixed muscular patterns are deviations from this ideal. They severely restrict one's options. A pattern based on hopelessness destroys ambition. A pattern based on fear undermines trust and corrupts the chances for close, warm contact. Each muscular pattern is associated with a particular underlying feeling, so the number of basic patterns is to some extent limited. There is no limit, however, to the shades and subtleties of their combinations. These underlying feelings and their interactions with the forces of growth, produce the infinite variety of personalities. It may be more accurate to speak of trends and proportions than to try to label people as one type or another. Yet each individual's body does speak more or less clearly of the patterns which possess it.

To see these patterns, and read the messages they contain, one needs a willingness to be affected by whatever is there. It is not only a matter of looking with the eyes; it is as much a sensing with the heart. It is a subtle sensing, of energies and vibrations not easily explainable. To the degree that we are open, unafraid, and consciously let ourselves be acted upon by another individual, to that degree we see his patterns, sense his energies, feel his pain, and know him. So: what the body reveals lies as much within us as within the someone we are trying to know.

In all of this there is a kind of unity. It follows from the fact that everything about a person originates from the core of his being. A person's body, his behavior, his personality, the way he moves, what he talks about, his attitudes, dreams, perceptions, posture, are all parts of a unitary whole. All are expressions of his core. They are interrelated; you cannot change one without influencing the others. Though they may seem at times to be independent, a definite theme runs through all of them.

To take an extreme example, the schizophrenic is a fragmented person. His speech is fragmented; his movements are awkward and uncoordinated; his actions are impulsive and very often incomplete; his body is tense in one place and flaccid in others; his breathing is grossly inefficient; his life and

dreams are filled with scattered images. He often feels unreal, like he was someone else watching his own actions, as if he were not in his own body. He seems distant, split off from those around him and out of touch with simple, physical realities. He has the vague eyes of a man preoccupied, and indeed he is. He is split on every level. It is this fact—that everything about him is in pieces—that is his unitary theme. His core struggles desperately against disintegration and just as desperately tries to put the pieces back together. His terror of being torn asunder by forces too great for him to contain destroys the possibility of smooth, coordinated efforts. He is sailing in a hurricane. He expresses this terrifying struggle in *every* aspect of his being. He has that kind of unity. So do we all.

For those who have preserved or regained their wholeness, impulse and expression flow easily, without the effort and struggle that characterize the troubled. This flow and wholeness is disrupted, for example, in those who feel anger, but never show it, or in those who cannot feel sexual attraction and love for the same person. In these people the flow of energy in the body is disrupted. (Breaks in the normally smooth curves of their bodies are an indication.)

We presume that such splitting begins early in life. Here is an example from our friend Sam Pasiencier:

"I was in a plane. There was a young couple with a baby, sitting across from me. The baby had those incredible, wide-open, infant eyes, and it was, just looking and looking. Every time the baby would reach out toward some object or another, the mother would take the baby's arm and put it back. So, right off, she was breaking the infant's unity."

If the mother continues to do that, several things could happen. The child could give up in defeat. Later, as an adult, his arms hang lifeless from drooping, narrow shoulders. He doesn't reach out to life. He waits for things to come to him. Or maybe the infant struggles and learns to get what he wants by being forceful. His jaw juts forward and his body becomes muscular and energetic. Things gently given elude him. Either way, the easy flow of curiosity and action has been destroyed. Spontaneity, which needs the broadest base of

open possibilities, has been lost. Feelings of defeat or conquest
have come to dominate.

One's whole style of life evolves from such interactions between mother and infant. And, in every case, the body is a
clear reflection of that evolution.

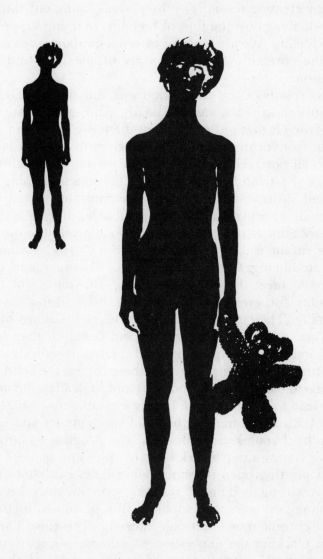

When the innate wholeness of the body is disrupted, we are banning from awareness the impulses that arise in our bellies, genitals, hearts, arms, legs, and other parts of the body. We are avoiding the feelings associated with those parts: anger and reaching out in the arms, sexuality in the genitals, love in the heart, and the full or empty feelings in the belly. We block them by creating tensions in those areas, using our muscles to throw dams against the flow of feelings. In doing so, we diminish all feeling. We are without internal guidance, out of touch with the core; life becomes empty of meaning and full of confusion.

Fears rob us of this connection with our deeper selves; central, important fears, such as death, pain, and isolation. The tragic irony is that in fleeing from them, we give them energy. To run from death, we deaden ourselves. We cut off fear and, with it, all our feelings. We create a pseudo self, with pseudo feelings. To avoid being left out we become something we are not, and so leave ourselves out. We constantly avoid what is irrational, unexplainable, or unjustifiable, and so lose touch with those things that simply *are*, the real ground of our being.

The infant is born with the capacity to be a whole and integrated being. Given the love, understanding, and support needed to meet the demands of growth, fullness of life is a possibility for every normal, healthy baby. But it's a fragile possibility. The child is easily overwhelmed. Callous or brutal treatment—threats, punishments, and demands that deny his needs, impulses, and perception of the truth—quickly destroys his natural spontaneity. To meet these forces, the child creates tensions to block his fear and pain and to deaden the impulses which lead to these feelings. Under tension, sensitivity is diminished. With needs unmet, the child compensates and searches for indirect routes to satisfaction. His progress toward adulthood becomes a patchwork quilt of roles and games.

In a healthy, open person, feeling flows easily into expression. A strong feeling of sadness spontaneously becomes a trembling jaw, tears, and sobbing. In a person with emotional blocks, chronic muscle tension interrupts this flow. For example, in blocking the expression of sadness, we tense the jaw, chest, stomach, diaphragm, and some muscles of the throat and face—all the areas which move spontaneously when the feeling is allowed its natural outlets. If the sadness is deep and long-standing, and the blocking continues, the tension will become habit and the capacity to express, frozen. With the growth of habit, awareness dims. The feeling itself may slip from consciousness, and situations which arouse it may be avoided. It is this habit and lack of awareness we are calling a block.

An individual may have several blocks of varying severity. These blocks as much as anything else give him or her their particular personality and character, and tell us who he or she is and how they got that way.

The pattern of muscle tensions in the blocks affects movement, posture, growth, and therefore structure. Changes occur in the skin tone and temperature. Blocks impede the normal flow of energy in the body. They impede not just chemical or mechanical energy, but that special life force which gives the others meaning. Beliefs, perceptions, and needs are the true energizers of human action. This kind of energy, this constant flow of feeling and purpose, is disrupted by blocks.

This energy is experienced in the body as streaming, and is dealt with directly in body-oriented therapies. The therapist presumes it, looks for it, and talks about it. If a person possesses a long history of energy being removed from a part of his body, that part will grow less. It may feel cold to the touch and will probably have less color. This energy can move very quickly, for instance, giving the eyes a sparkle or giving the face a glow. It can also move inward toward the core, leaving the outside cool as in shock or withdrawal. Biological energy can stream through the body freely or it can be blocked in the body and not be felt at all. Or it can block in certain areas and flow into others. For example, with excess energy streaming into the head, you're going to have a lot of thinking; or, if it's really excessive, you can produce degrees of mental excitation right up to psychosis.

The process of undoing blocks involves arduous, sometimes painful, persistent work. The insidious, interlocking nature of fearful attitudes, habitual muscle tensions, blocked feelings and restricted awareness makes any change both difficult and delicate. Once the process of change is begun, however, momentum in the direction of growth can increase to the point where efforts that once seemed impossible become automatic, and actions once approached fearfully are taken in stride. For most ordinary people beset with problems, there's a way out if only they will take it. Courage, faith, hard work, wise counsel, and truth are the essentials.

The work done must activate changes on physical, emotional, and mental levels, since each level affects the others. We feel better when we're physically healthy, just as we have a better chance of staying physically healthy if we're emotionally content. The physical work in undoing blocks may involve systematic exercises like yoga or bioenergetics, a proper diet, active intervention like rolfing, màssage, chiropractics, exercises to increase awareness and change habits of posture and movement like Judith Aston's patterning, Feldenkrais's work, or the Alexander technique. This work is basic. It builds the capacity to handle deep emotional changes and release the energy needed to follow them through.

Deep change is much more than finding a comfortable role or playing better games. It is a true expansion of the self, a removal of self-imposed limits—restrictions grounded in irrational fears and childhood defeats. These fears must be contacted and reexperienced. The attitudes to which they give life must be brought to awareness, then examined, and the whole process enlightened with persistent self-discovery. A new base must be built up on physical vitality, realistic attitudes, emotional satisfaction, and the acceptance of life. By pursuing growth, by going deeper and deeper into our feelings, by seeking within ourselves the source and meaning of our lives, we can only come to find an unending spiritual reservoir—ineffable, mysterious, and yet the surest, truest ground of our existence.

It is wise to remember that a person's patterns always contain pain and fear. They are intimate, and the embodiments of much suffering. Skill is required and compassion is essential if one is going to make contact with them and help dissolve them. A long time in the making, they do not yield easily. Force does not work, but tenderness, respect, loving understanding, and a commitment to be honest will often be enough. Strength and courage are needed to break free. For, ultimately, these patterns are bonds which imprison man's spirit. They bind us to self-concern and painfully isolate us from each other. It is first by seeing them and then understanding them that we can best begin to free each other from their grip.

2.

A CONVERSATION

This is a transcript of one part of a long conversation we had during the writing of the next to last chapter of the book, "The Authors Look at Themselves." We include it here to give the reader some of the flavor of the body-oriented therapists' ideas. Our friend Sam Pasiencier is the third voice.

Sam: Functional expression, I have an image of this. I see a child walking to a cupboard, feeling hungry, taking out something to eat, and eating it. That's direct. It's an immediate expression of organismic needs. It involves the whole body, the legs, arms, eyes, the feeling of hunger in the stomach, and so on. You can also see a child who is not hungry, who wants attention from his mother. He acts as if he's hungry and may even believe that he is hungry. He goes to his mother. He doesn't go to the cookie jar. He says, "I'm hungry. I want this or that." There is a deviousness in that. It's not the direct functional expression of the need. It's like scratching in the wrong place. You're itching in your left eye and scratching your right.

Hector: So functional is what meets your real organismic needs. Whereas, dysfunctional would be those behaviors that meet substitute needs. That's nice.

Ron: And the body structure will reflect that dysfunctional behavior. You can see that years later a person with that attention thing may put on a lot of weight by eating when he really wants human contact.

H: This seems to be our fundamental premise, that the body reflects feelings, attitudes, perceptions of the world, and oneself.

S: And it does it through its function. You can see the body like a stalactite in a cave—the history of the total function of the organism.

H: Structure and function in relationship. What is functional is functional if it meets the real needs of the organism. Game playing, so-called neurotic game playing is functional in another way. It is an attempt to fulfill substitute needs.

R: It usually is unsuccessful.

S: It's doomed to be unsuccessful, because the person doesn't know his real motives.

H: Repeating the dysfunctional thing keeps the person split.

S: And you're always under tension because the real need is never satisfied.

R: Right. The game is designed to provide satisfaction, ego satisfaction, not organismic satisfaction. For example, having sex to prove your potency. It might prove that, but it might also fail to provide any real sexual release.

S: That's the thing. It doesn't give any release for honest organismic needs. The tension is still there.

H: So, game playing is dysfunctional. It reinforces splitting.

S: That's what drives other people crazy. If you're in a relationship with a game player, there's no way to make them happy, they're not expressing their real needs, and it'll drive you out of your mind.

H: Therefore, our premise is that persistent game playing produces game playing changes in the body that are visible and are cues, once you get to know that language.

R: A simple example is locking up anger in the shoulders in order to avoid alienating people.

S: If you want to, you can use almost anything as a hammer. But, whatever that thing is, if it's not a hammer, you're liable to screw it up using it as if it were. You can use the wrong tool to do a job, but you mess up the tool and you change its form in the process.

R: Yes, walking can be seriously hampered or distorted when the body is also being used to block sexual feelings at the same time. The structure changes under the impact of that kind of block. Some muscles that should be helping stop participating in the walking and the others have to take over. When this become habit, the structure distorts. Or let's say a child has an impulse to reach out to its mother and receives a hateful look. It now also has an impulse to shrink in fear. The child can't live with those two impulses. The result is conflict.

H: And the body response to this conflict is to block one or the other of the feelings involved, in this case, the fear or the longing. It might block the fear by pulling up the shoulders and tightening the throat. This suppresses the fear, by freezing it into a partial scream, an incompleted response that keeps the fear and even the perception of the mother's hatred outside awareness. That allows the longing for her to complete itself in continued reaching out. The child might as easily have blocked the longing by deadening his arms, his belly and his shoulders. The psychological and physical solutions are like shadows of each other.

R: The psychological events can't be separated from the body's response.

H: So a Freudian is going to look at it on the psychic level. He's going to try to work out the psychic mechanisms and decondition the person that way. He sees the neurosis that way. He listens to what the person is saying. We see the neurosis by looking at the person.

R: Well, since there's unity, you can look at anything about the person and have some chance of picking it up.

H: But here, in terms of our task, it's the physical structure. The physical structure demonstrates the mechanisms of compensation which attempt to achieve the best functional capability. 'Cause they are all attempts to do the best possible.

S: This leads us toward a thing about awareness, too, because in order to block, you have to cut feeling in the area.

For a while, I guess, the child feels that it's blocking. But then, he even forgets that he is blocking and gets to repression, and he removes his awareness from the whole area. It becomes numb and the block, I think, becomes continuous.

H: So we're saying that the blocks that appear in the body, in the physical body, are most often repressed, they are not in awareness, they are not in consciousness. The person goes with his shoulders pulled back and he's not aware of this.

R: If you ask him to relax, the shoulders stay pulled back.

H: You point it out to him and he says, "Well, I've never seen that before. What is that?" And still, he can't let them go.

R: For any person, his body is a solution to his conflicts, and you can't take that away from him. Until he's really ready for change, he's gonna hold on for dear life.

3.

BASICS

Man is an animal who is split halfway up the middle and walks on the split end.

—Ogden Nash

"Energy and gravity" are two fundamental terms in modern physics. They are irreducible, given and unexplainable, as yet, by any more basic concepts. We turn to these terms for a beginning. Along with these, we discuss some others, such as "grounding" and "core" and "extrinsics" in order to give roots to the ideas put forward throughout this book.

Man is an upright animal. He lives in constant intimacy with the force of gravity and the manifold variations of energy. His relationship to these forces are the most central of his existence. His ability to contact basic, physical, emotional, and spiritual realities and the means whereby he does so are at the core of any system with which he tries to understand himself.

ENERGY

The land is a mother that never dies.
—A Maori Saying

Our organisms interact with various energies. Apart from sunlight and the energy in the foods we eat, there are the electromagnetic currents around the earth to the poles, gravity's pull to the core, the spin of the planet in its orbit around the sun, the pull of the moon, as well as that of the planets.

It is well known that animals sleep and hunt in relation to

these waves of energy. Fish in the oceans feed in rhythm with the tides. These same forces influence our lives at every moment. We interact with these forces through our skin, which is sensitive to touch, density, vibration, heat, and cold. Our eyes see light, our ears hear sound, our noses smell and tongues taste. A mechanism in our inner ear constantly senses our relation to gravity.

These are energies known to us through our senses or described by our contemporary scientific equipment. Certain other energies—we can call them subtle energies—have been described throughout the history of man. Wilhelm Reich reported that he could accumulate a special life-giving energy in a device called an orgone accumulator. The great pyramids of Egypt may have served a similar function. Mystics all over the world have been and are working to open themselves up to these energies, known by various names: prana* and cosmic energy are but two. Consciousness of these phenomena as an available experience is being described by more and more people.** When we are able to align ourselves physically and emotionally, we open a channel to receive this energy and, in so doing, we become awakened on a level hardly to be compared with ordinary consciousness. This heightened consciousness is not simply intellectual; it involves our very tissues.

To experience such states, one must be able to permit change. A certain amount of fluidity is necessary. The Yogis of the East describe experiences in which energy that feels like a warm liquid streams up the spinal column, filling the brain, and allowing consciousness to expand from a limited "I" to a direct awareness of the deepest spiritual truths. "Streaming" is also reported by people in Reichian and bioenergetic therapy, transcendental meditation and other therapeutic and religious pathways. This warm, fluid streaming can be contrasted with the lack of freedom, emotionally and physically, that is most

* Prana refers to a subtle energy infused within the air we breath which the Yogis of the East attempt to absorb within the body by various physical and meditative techniques, such as Hatha Yoga and Kundalini Yoga.

** See *The Evolutionary Energy of the Kundalini*, Gopi Krishna. The Shambhala Press.

often described as a lack of motion: being stuck, frozen, blocked, held, trapped; in general, stasis.

When the energy that is available to give life and vitality to a person does not flow, stasis results with a jamming (confusing overactivity) in the central nervous system. This jamming is manifest as "chatter" in the mind. The musculature responds by "holding" or blocking flow. The more internal chatter we have, the less external input our nervous apparatus is able to receive and act upon. This chatter is repetitious and habitual, as in a repeating tape loop.* The same themes, attitudes, problems and solutions appear over and over again. In the face of this repetition, we seem helpless. These deeply ingrained habits of thought and feeling have been produced by repeated life experiences, often originating in our earliest years. As our strongest habits, they tend to dominate our behavior and govern our immediate responses to almost all situations. For example, if our parents kept giving us a mixed love/hate message, leaving us uncertain and insecure, then in later life our deepest interpersonal relations are sure to include this uncertainty. Not certain of being loved, we keep asking, "Do you love me? Do you love me? Do you love me?" either directly or through our actions. With our nervous apparatus continuously locked into this struggle, we are unable to perceive other, more nourishing energies, or to free ourselves of doubt.

This uncertainty will inevitably be translated into a body statement. The individual's every gesture will be a statement seeking validation. His eyes will search you out for approval. He will move toward you tentatively. Indeed, the body has no choice. It displays the total dynamics of the individual. The circuitry of the nervous system, when so organized, restricts and contains the available options for response. To this extent we are preprogrammed. In esoteric schools, this is often referred to as a crystalization of the personality.

---

* Such chatter can be contacted by lying quietly and simply "listening" to the activity of the mind. For most of us, quite a circus of thoughts, sounds, ideas, memories, images, and colorful patterns is going on.

Other mechanisms producing chatter are states of pain, exhaustion and intense stress. We are all aware of how a headache or a throbbing toothache cuts out all awareness except for itself. On a more subtle level, chronic low-grade pain such as a neck ache or common low back pain may send constant impulses into the nervous system which, by taking up space and time limit the number of other events (external and internal) which can be processed. Exhaustion is a state in which so much input has occurred that the body energies are in a sense soaked up in the tissues. They simply have had too much. They need time to settle and gradually discharge. Here again, our nervous system is committed to a priority in its attention, namely, relieving fatigue. It simply is materially unable to respond further.

The number of situations in our culture capable of producing such overload is increasing. We are inundated by lights and sounds: TV, radio, motors, subway crowds, advertisements, movies; not to mention the myriad number of electromagnetic waves of every kind filling the air: radar, radio communications, telephone lines. Larger planetary disturbances are produced by atomic testing and fallout, earthquakes, sunspots, and electrical storms. Any of these wave forms, once large enough, can disrupt our energy fields.

In light of all the above, we may define chatter, in terms of the central nervous system, as any ongoing process which compels attention to itself, limits regenerating inputs and consumes energy, thereby reducing the organism's ability to respond to its environment and expand. The injunction of almost every school aimed at increasing personal growth and consciousness is, "Still the mind!" This stillness allows our internal energies to harmonize with those available in the vast, potentially nourishing external cosmic pool.

The other significant factor we mentioned in limiting expansion and filling in the organism is termed "holding." Here, we are concerned with interference with the energy flow through the body. What is this flow? When is a body open and freely streaming energy? Central to all life forms appears to be the capacity to expand and contract, to pulsate. We have within us some systems that demonstrate this. Our circulatory system expands and contracts with every heartbeat. Our blood vessels fill as our heart contracts, and contract as our heart dilates. So we have a dance of filling and emptying between our heart and vessels. So great is this wave of flux that, if you lie down quietly on a bed (waterbeds are particularly well suited), it is clearly felt. Certainly, we have all felt pounding throughout our body after strenuous physical exertion.

The lungs are a second great energy pump. The breath of life is a wave of air, flowing in and out, in a cycle of expansion and contraction. The body also appears to possess other complex pulsatory cycles which have to do with such basic life functions as eating and sleeping. In the field of acupuncture, it is known that the internal organs have periods of waxing and waning, both on a daily and seasonal basis. Certainly, animals have life cycles of reproduction, feeding and rest which are related to larger, biophysical, planetary pulses. If we imagine the body as a nucleus around which a complex series of pulsating waves weaves a pattern of balanced periods and harmonics, we begin to approximate the wave dynamics of life. The following sketch, adapted from a concept by Dr. Randolf Stone, suggests and illustrates this point.

As we've said, these pulses include those originating within, like heartbeat and respiration, and wider seasonal and daily rhythmic fluxes. In fact, only as we harmonize our internal pulses with those in our environment do we experience full vitality. In working with people, it has been our experience that when internal and external wave patterns syncronize, a fundamental pulsation occurs. This pulse starts deep in the core or center of the body and spreads outward toward the periphery. This is accompanied by a definite glow surrounding the body. The skin becomes warm and literally vibrant, the eyes bright and soft, the respiration smooth and full, allowing both chest and abdomen unguarded expansion. This is the ideal. Unfortunately, it is rare to see anyone living in our technical/industrial society who exhibits this degree of alive-

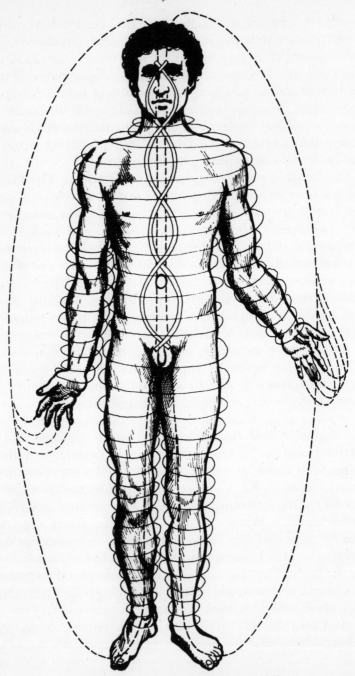

FIGURE 1. WAVES OF ENERGY IN MAN.

ness. With the number of family and work demands placed upon us, in the limited space of crowded cities, survival takes precedence. What this means is that from early childhood, we begin to learn that holding back compromising feeling is safer than allowing free synchronous pulsing. This holding is most often manifest as rings of muscle and fascial tension in areas between the major segments of the body. These areas are the neck and upper shoulders, the diaphragm, which is the juncture of the chest and abdomen, the lower back between the abdomen and pelvis, the groins, separating thighs from trunk, the knees and ankles. The feet and eyes can also be held.

These are common areas of complaint. The pulsing waves of vital energy within us keep striking these muscle blocks, augmenting the charge and tension within them. These rings of tension involve the body from its outermost surface to its deepest core, and the amount of energy passing from segment to segment can be so reduced that gross changes in form, color, and development occur. Hands and feet may present themselves as cold and small. The head may be large and congested, or the belly blown up while the chest is collapsed. (These aspects of holding are presented later in more detail.) In general, in these areas of increased muscle tension, blood supply is reduced. This leads to the collection of tissue wastes, setting up a mechanism of toxic spasm and stasis. (Tissue wastes cause this by changing the local oxygen and acid content of the tissues.) To this, the nervous system responds by firing more and more signals, increasing the local charge. Anyone who has had a backache or heartache knows of this relation of pain-spasm-pain-spasm. Eventually, if this occurs for a long enough time, a physical metamorphosis occurs. The tissues harden in an attempt to splint the area against further insult, producing a fixed structural block.

A backache may be produced by a slipped disc, but more often it is the expression of our struggle to hold ourselves up or back. A heartache is often the expression of the pain of our blocked impulses to love or be loved. We are afraid to pulse freely. Anxiety, alienation, and an ever-increasing sense of inner despair can be recognized in those around us as well as

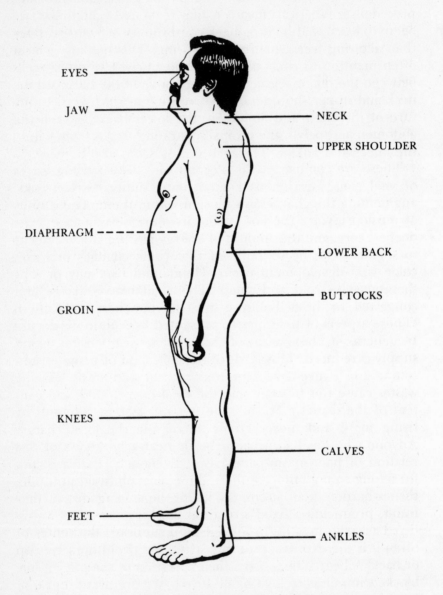

FIGURE 2. MAJOR AREAS OF HOLDING.

felt within ourselves. As these provoke ever-tightening tensions and blocks, we separate farther from the pulse of life. Our inner flow becomes a trickle behind a gray dam of blocked emotion. Love, which is the key to transcending these blocks, is frightening to us. Who indeed possesses the courage to allow an uncontained discharge of vitality to occur? The fear of drowning in our own emotion is simply too great.

Unless guided by someone who recognizes the need to unfold this inner charge with love and patience, we are trapped in a vicious circle of tension, insecurity and further tension. In such a vicious circle, depressed vitality must result. So, in effect, without help, the areas of block continue increasing in charge until an explosion occurs manifesting itself as rage, irrational anger, or other forms of violent action. Can the mounting number of rape and murder victims be an expression of this? A very convincing case could be made for the idea that the current epidemic of heart attacks in this country is related to an inner displacement of this blocked energy. The displacement may be so marked that the entire body becomes more and more held, leading to an inability to move except in rigid patterns. These blocks are far from subtle. They are obvious and extensive. They can be easily detected by the application of pressure with the fingers.

GRAVITY

*Gravity is the root of
all grace.*
   –Lao Tzu

Gravity acts upon us every moment of our existence: sitting, standing, during all movement, and even lying down. This force, whose action we can describe but whose real nature we do not understand, pulls us unceasingly toward the center of the earth. Generally, we stand our ground in constant conflict with it, but it is just this encounter—standing upright against gravity—which gives us our ability to reach upward and outward. Harmony with gravity aids us in this endeavor; disharmony plunges us into an endless struggle, demanding large amounts of our available strength just to stand up, making it much more difficult to face a demanding world.

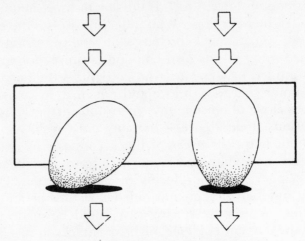

FIGURE 3. GRAVITY AND ALIGNMENT.

Abstractly, the general shape of the body is that of an elongated egg. Seeing it as such, it becomes clear that maintaining an erect stance can be a most difficult task. As is evident in the next figure, proper alignment actually uses gravity to hold the egg up. With a flattened base, the egg is actually made more stable by gravity's pull downward. But once alignment deviates enough, gravity acts to topple it. It is the same with the human body. If a part of our bodies is off center, we are forced to expend considerable energy just to hold ourselves up. In such cases, the body then realigns so as to maintain a compensated balance, often a precarious one. For example, if the chest is going one way, the belly will go another (as can be seen in Figure 4).

In viewing a body, one of the most important observations to make has to do with its relation to gravity. Does the head sit over the shoulders? Does the chest fit over the lower half of the torso? Is the pelvis supporting the large segments above it? Are the legs and feet under the body?

FIGURE 4. COMPENSATED BALANCE.

FIGURE 5. THE LATERAL LINE.

The person depicted in Figure 5 approximates, but does not represent, the ideal lateral line.

The ideal axis for obtaining the greatest balance is that which connects points at the top of the head, middle of the ear, middle of the shoulder, midpoint of the hip joint, center of the knee joint, and center of the ankle joint. This line will also pass through the juncture of the lower backbones and the large triangular bone sitting at the base of the spine (sacrum). When these points are so aligned, each underlying segment supports those above it. Like the on-center egg, gravity then keeps the body rooted, and a minimum expenditure of energy is required to maintain an upright stance. Considering that our muscles make up a large part of our total body weight, and consume a major portion of our body fuel, it is easy to see that their efficient functioning is of central importance. If we consume unnecessary amounts of energy to stand and move, we deplete the supply available for other activities. The more energy consumed in holding ourselves up, the greater the amount of chatter in our internal and projected energy fields. This reduces our vital pulsation, resulting in diminished flow with our environment.

When we are out of balance, our every move is weighted down by gravity. With balance, the ground is a pillar of support, and gravity our ally. Emotionally, we cannot help but be influenced by this relationship. Our nervous system constantly endeavors to come to terms with messages from our muscles (not usually conscious), informing us that we are being pushed down or weighted down. Many people who express feelings of being overburdened have bodies that are bent forward toward the ground. Others, whose bodies are bent backward, resisting and pushing against gravity, experience life as an unending struggle.

Of course, their experience correctly reflects what is happening. Bowed forward, they are literally carrying the weight of the world on their shoulders. Bowed backward, they are pushing upward against a force so great that most of their vitality is involved in this effort. When the lines of the body are in

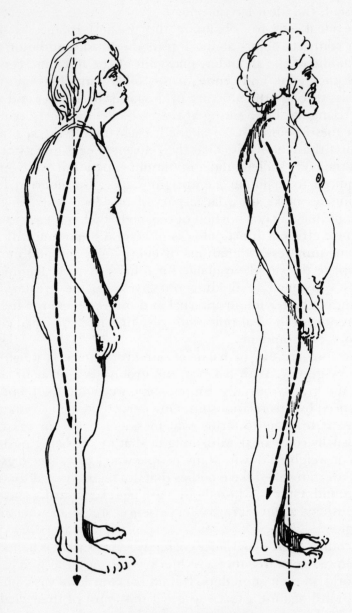

FIGURE 6. MISALIGNMENT.

harmony with gravity, the various radiant forces available in the environment flow through us freely. There is no resistance.* These lines are like conduits or rivers of energy around which fields develop, streaming vitality into our tissues. It is significant to note that these are straight lines. (Recall Figure 5.) As they bend, their capacity to transmit energy decreases.

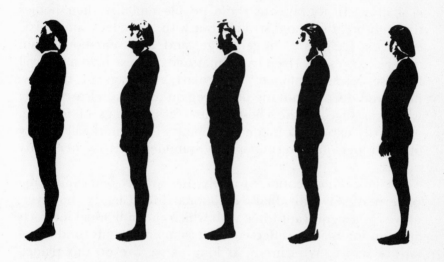

FIGURE 7. THE STRUGGLE WITH GRAVITY.

All of the people in Figure 7 are involved in a struggle with gravity. Each body presents a different degree of backward bowing. All are shorter along the back than the front. With significant bowing, these people will each display attitudes that can best be described as rigid. They tend to be strongly determined in their directions, with fixed ideas about right and wrong. Often pushing toward what they think they want, they contain their feelings, especially those of tenderness and those

* The situation is similar to that of a magnetic needle in a magnetic field. If the needle is able to swing freely, it will align itself with the field and just sit there. No further work is required.

of hurt. They have reached a point in their psychosexual development where their drive is normally directed but limited in full expression. The head is thrown back, the shoulders are pulled back, the long muscles of the back (spinal muscles), which travel from the base of the spine to the base of the skull, are short and tight. The pelvis is cocked backward and held, unable to swing forward and attain full sexual release. Respiration is held in, with a very limited excursion so that breathing is shallow. In therapy, as these people mobilize their bodies through breathing and striking out with arms, legs, and pelvic exercises, feelings of anger, rage, and deep emotional pain may be expressed. Deep tissue massage of these held areas will likewise release emotion. Particularly, massage of the jaw, neck, and deep within the pelvis region tends to release anger; while massage of the chest often frees feelings of longing, outwardly expressed by deep sobbing. The emotional history of each person determines the statement his or her body makes.

It is not our intention to imply that gravity determines the pattern of an individual's emotional life. Clearly, however, gravity is a significant force with which the individual interacts and this interaction reflects as well as maintains his fundamental life stance. We cannot, as long as we live on this planet, escape it.

CORE AND EXTRINSICS

In this section, we will make a very simple distinction between two levels of experience, and relate these to the body.

The body, as we shall see, can be visualized as made up of an inner, or core, layer and an outer, or extrinsic, layer. In our conceptualization, the core layer is involved with "beingness," and the extrinsic layer with action or doing. Beingness refers to that part of us which exists independently of structured identifications. It has the qualities of spontaneity, direct connection to organic realities (experienced as such), a sense of

effortlessness and immediacy. Contact with this level of experience is most often refreshing and revitalizing.

Without any labels, we still "are." Without being busy, doing
anything or even wanting anything, we still exist and experience our existence. Beingness, when developed, gives us the
capacity to make contact not only with each other and physical
reality, but with all that is and is not. It is a "place inside" to
which we may go for sustenance. Dropping our roles and
titles, our claims upon the course of events, reaching within
ourselves, quietly allowing whatever is there to surface, we
begin to contact this inner self.

Contrasted with this being layer is the outer, extrinsic layer
which is concerned with doing, taking action, carrying out
plans. It is associated with the experiences of deliberateness,
effort, will, and desire, with maneuvers, postures, set attitudes,
roles. It affords less contact, by virtue of limiting and directing
the flow of events through focussed expectations.

In terms of the body's structure, the core finds part of its
physical home in the small, intrinsic muscles linking the various segments of the spine.* These fine muscles coordinate
delicate movements in time and space. When functioning
properly, they establish a center around which larger movements occur. The energy flowing from the core modulates
these larger movements. Balanced with these core muscles are
the extrinsic muscles which are involved in actions requiring
strength. The extrinsics include such muscles as those of the
arms and legs. Without the internal balance given by the core,
the actions carried out by the extrinsics lack flow, grace and
harmony. The extreme of this would be a person who could
be called "musclebound," possessing extrinsic strength while
lacking core. Such a person has an imbalance in favor of the
extrinsics and his movements, consequently, lack grace. When

---

* The intrinsic muscles include not only those joining spinal segments, but also the
small muscles joining the small bones of the hands and feet, as well as those deep
muscles joining the head (base of the skull) to the cervical spine. In each case, they are
involved with movements of great refinement.

our core structure lacks organization, or has never been fully developed, the body compensates by employing the extrinsics. This is a chronic, unremitting situation. The development of the core takes years, and usually occurs quite normally during early childhood. The postural and emotional habits that result when this development is disturbed also take years to evolve. They grow to be extremely persistent through years of continual repetition. Such compensations can be clearly seen in situations where a force impinges on us, a force such as gravity or the emotional demands of living in the world. Use of the extrinsic musculature is an attempt "to hold up against it all." In some people, even the extrinsics are not adequately developed. In their bodily and psychological attitudes, they seem to be asking the world about them to hold them up. These people were originally called "oral" types.*

Simply put, they lack sufficient development of their internal and external energy layers. Their inner experience is one of weakness and debility. Others, as we have seen, attempt to

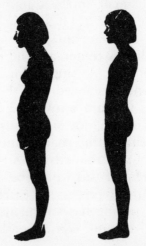

FIGURE 8. ORAL AND RIGID TYPE CONTRASTED.

* See "Character Analysis," Wilhelm Reich; and "The Language of the Body," Alexander Lowen.

hold themselves up by employing their extrinsic musculature, using a continual contraction of muscles which would normally be more relaxed. These latter are referred to as "rigid" types. Both rigid and oral lack a strong inner core or center. As long as this is the case, their every action and compensatory reaction will be unbalanced. Most of this process goes on outside of awareness. In fact, in order to change, the first step must be to become conscious of this inner disharmony. When that happens, energy begins feeding into the deeper core, creating the possibility of wholeness. As the core becomes vitalized, the individual begins to move from his center. His body and his attitudes change. His emotional dependency and the constrictions of defensive attitudes yield to a sense of self, and open, flexible interchange with others. He finds he no longer needs the external support or extrinsic rigidity to hold himself up. He can surrender these, and begin to feel the pleasure of release and the solidity of an integrated self.

ANATOMY OF CORE AND EXTRINSIC MUSCULATURE

The physical structures that make up the core include the following:

1. The bone marrow.

2. The bones themselves, particularly the spinal column and long bones.

3. The fascial sheaths of tissue that condense close to bones into bands known as ligaments. (Note: these include the long bands of tissue which wrap about the spine from the base of the skull to the base of the spine.)

4. The already mentioned intrinsic muscles (small muscles linking the bones with the spine).

5. The muscles which go from rib to rib (intercostals).

6. The diaphragm.

7. A big muscle (psoas) which goes from deep inside the body; in fact, from the spine outside onto the large bone (femur) of the thigh.

The heart and great blood vessels are also part of our core being. Our language includes many expressions which point to this connection. "My heart cries for you." He or she is "heartbroken." We recognize when someone is "speaking from the heart." And we talk of "our deepest, most heartfelt feelings." Similarly, the blood itself contains a deep sense of our most inner being. We speak of "blood brothers," or a "blood tie." We use phrases such as "That makes my blood boil," or "hot- or cold-blooded." It is here that the relation between the core and our emotional life is most clearly evident. On the deepest level, our emotionality is expressed in the beat of our heart. It beats out the rhythm of our innermost being. Fear, love, panic, and joy are all experienced as heartfelt emotions.

The outer, or extrinsic, layer of the body consists of the large muscles involved in maintaining an erect posture, and the layers of tissue enveloping them. These are:

1. A very broad muscle which covers two-thirds of the back and connects the pelvis, trunk, and arms (lattisimus dorsi).

2. The long muscles outside of the spine, involved in keeping the spine erect (sacro spinalis).

3. The large muscles of the thighs and arms.

4. The large muscles which make up the major portion of the front of the trunk, from lower chest to include the entire front of the abdomen (belly wall).

5. The large muscle going from the upper two-thirds of the thorax to the arms (pectoralis major).

6. The tissue sheaths surrounding these muscles (fascia).

7. The fatty tissue under the skin.

8. The skin itself.

In a simplified, schematic image of the preceding anatomy, the body may be viewed as consisting of inner and outer cylinders. At birth, the inner cylinder or core is present, though underdeveloped. If the environment is appropriate, the individual develops a solid core. On the other hand, if the individual's emotional or spiritual development is blocked or distorted, the core is experienced as lacking or inadequate.

Seen in structural terms, the core is in the internal support-ing pillar. It is important to understand that the outer layer develops in relation to the growth of the core. When the core is firm, the outer layer builds up about the core without strain. When it is weak, the outer layer either does not develop, and we have an undernourished, underdeveloped, weak indi-vidual, or in various ways the outer layer overdevelops in an attempt to hold the structure up, or together. In this type, the compensatory relationship of holding and core development is easily recognized.

The amount of chatter present in the central nervous sys-tem is directly proportional to the degree of physical holding and, therefore, blocked emotional and spiritual development. It follows that the more impaired the integrity of the core, the greater the volume of chatter. Naturally, an interplay occurs in which chatter and low-core integrity create holding, which creates chatter, which in turn masks input to the core. The same interdependence also exists on the emotional interper-sonal level. A low sense of self is, for example, often compen-sated by constant self reference, seemingly endless reports of one's activities, and bragging. All these lead away from mean-ingful contact with one's inner feelings and, therefore, toward basically irrelevant communication. Little meaningful interac-tion takes place, and very little emotional nourishment is re-ceived, leaving the self unexpressed, unfulfilled, and still de-pleted.

So the greater the amount of holding, the less the amount of core. Another way of stating the above is: in an individual's development, the less energy input arriving as love and secu-rity into the central core being, the greater the chatter and holding in the total body/mind. Chatter significantly affects the individual's mental state. Little chatter is accompanied by clear thinking, serenity, and balanced responses; increased chatter with rigidity, confusion, and imbalance. Holding ex-presses itself in one's physical structure. Although we have conceptualized that chatter affects mental state and holding, the physical structure, we must remember that this is an arbi-trary division, and in truth chatter and holding are insepar-ably intertwined in the reality of our living bodies.

GROUNDING

"Tie your camel to a tree,
then trust in God."
–The Prophet

The health of our core determines how we relate to the ground. If, while standing, we are able to relax and let ourselves fully contact the energy field of the ground beneath us, we experience a sense of solidity. We are "grounded," in contact with the reality of our being in time and space. We know, and for that moment accept, "where we stand." In contrast, many individuals stubbornly resist pressures best handled by letting go. They remain in unbearable work or love relationships despite recurring destructiveness. They are unremitting. Such an individual is stuck and chronically in pain. His contact with the ground and reality is so uncertain that letting go would plunge him into the panic of not knowing how to orient. Hence, in desperation, he elects to hold on. There is almost no limit to the pain he is willing to endure. We see him physically bent, obviously overburdened, with broad, rounded, hunched shoulders, head drooped onto his chest, his pelvis and thighs immobilized, his knees rigidly locked and his legs and feet tightly contracted.

The tension and stiffness in the lower half of the body, especially in the legs and feet, greatly reduce the sensitivity in those areas. The loss makes balancing precarious and floods the nervous system with chatter. It is as if the person were standing and walking on stilts. Instead of relaxing and regaining contact and balance, the chatter, which is basically a fear of falling, elicits a response of further tightening and bracing against the fall. Emotionally, this falling could be experienced as being taken advantage of (being a "pushover"), losing status (falling on one's face), not performing well enough (as in falling down on the job). It is no accident that in cultures which embrace a stiff formality, one walks gingerly: a loss of face is considered adequate reason for self-destruction.

When one's footing is tenuous, any change creates disorien-

FIGURE 9. AN OVERBURDENED INDIVIDUAL.

tation. If, as happens during sexual intercourse, the amount of energy streaming through the body is increased, the ungrounded individual often experiences this as a force which can be overwhelming. For such an individual, the only way to deal with these high energy drives is to contain them, thereby cutting off vitality. The pelvis, which contains a large reservoir of vital energy, is immobilized. (Generally, whenever there is a chronic lack of contact with the ground, the mobility of the pelvis is impaired.) Contact with the ground is essential for nondestructive, full streaming of energy through the body.

When the core is even more fragmented, and contact with the ground grossly uncertain, almost any energy input threatens to tear the individual apart. He or she lives with the persistent terror of becoming completely disorganized. Fear is enough to immobilize almost any attempt at self-assertion. To mobilize and move with flow would mean allowing free energy through the body, and this is more energy than can structurally or emotionally be handled. To avoid falling apart, he or she blocks flow by locking all his joints. Hence, movements are rigid, broken, and without flow from one segment of the body to the other. The head is usually tilted to one side and eyes are vacant and distant. Moving like a puppet on a string and looking like nothing so much as a broken doll, any vitality is experienced as an almost demonic force threatening to blow the self to pieces. He or she has all life's survival mechanisms employed in holding all the parts together. In most other situations of a core-ground insecurity, the level of reactive holding on the part of the outer muscular layer is such that the individual is able to confront life's demands. His compensations exhibit enough energy to avoid fragmenting easily. But, in severe cases, any stress is poorly withstood. Unable to deal with increased pressures from without or within, the severely fragmented individual will often totally cut himself off from his feelings. Such a cutoff reduces the energy flow in the body, leading to a state of physical and emotional detachment. As this becomes more and more extreme, the body/mind split reaches a point of almost total dissociation, loss of contact with reality, and, sometimes, complete immobility.

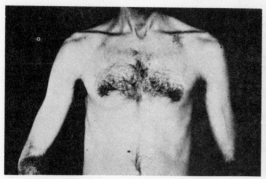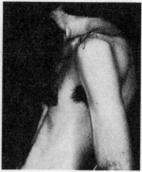

FIGURE 10. HOLDING AGAINST THE THREAT OF DISINTEGRATION.

In the preceding chapters we have looked at some of the essential aspects of the responses individuals make to energy, gravity, and the ground, and some of the ways development of the core influences these responses. We have described but a few of these patterned relationships, hoping only to introduce the reader to the reality of the body's adaptations to energetic and emotional life processes. In the next chapter, we look at the body in more detail.

THE SILKEN TENT

She is as in a field a silken tent
At midday when a sunny summer breeze
Has dried the dew and all its ropes relent,
So that in guys it gently sways at ease,
And its supporting central cedar pole,
That is its pinnacle to heavenward
And signifies the sureness of the soul,
Seems to owe naught to any single cord,
But strictly held by none, is loosely bound
By countless silken ties of love and thought
To everything on earth the compass round,
And only by one's slightly going taut
In the capriciousness of summer air
Is of the slightest bondage made aware.

—Robert Frost

4.

BODY PARTS

Like any other complex system, man is a composite of many parts, each displaying the dual nature of independent attributes, as well as a greater totality. Though it is the configuration of the whole which makes each man unique, one can profitably study the contribution of the separate pieces of that overall picture. In this section, we do just that. We look at various parts of the whole man: legs, face, feet, and other body parts. Each part, due to its unique place in the functioning of the organism, reveals something different about our total structural, mental, and emotional patterns.

THE OVERALL BODY

The general properties of structure that we notice in all people; thinness, heaviness, muscularity, maturity, all have psychological significance. The woman with a girl's body may very well have a strong wish to remain a child. She may even have consciously decided to do just that at an early age. The tall, thin body which seems to need nourishment may belong to a person who feels emotionally starved, while a heavy, sluggish body belongs to someone whose life is dull and painfully stagnant. In any case, the total body yields a general impression that can often be quite accurate. There are two features, however, of the overall body which we want to discuss in detail; displacements and asymmetries.

The first things to look at are large, segmental displacements. In these cases, the person is top- or bottom-heavy. The energy has moved either upward or downward on a large scale.

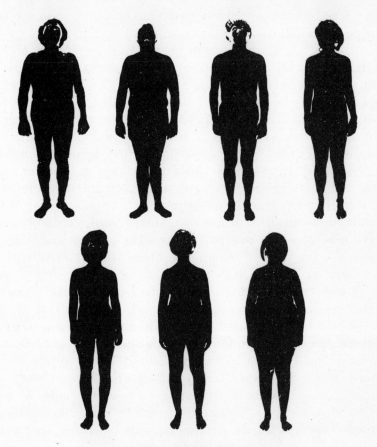

FIGURE 11. DISPLACEMENT.

In Figure 11, we have depicted a progression from the person on the left, who is overblown in his top half, through the fourth figure, who has good balance between top and bottom, to the last figure, whose bottom half is overblown. In bodies with large displacements, energy is not moving evenly through the entire length of the body. The main area of cutoff or block is at the level of the midback. The overblown segments (top or bottom) have collected a large amount of stagnant energy. There has been damming up, with a large lake of this static energy building up behind the dam. On the other side of the dam, the energy is just trickling by. These areas are less well developed and are often found to be disproportionately 'thin, with poor color, decreased blood flow, and lower skin temperature. In extreme cases, top and bottom halves can be so different that they seem to belong to different people.

Individuals showing extreme displacement are unable to contact major areas of feeling and vitality. Their bodies tell us at a glance of the inner struggle taking place. They move and live deeply divided, experiencing constriction as an unremitting constraint. As Figure 11 would suggest, displacement in men is most often in the upward direction, while in women it is downward. In each, the functions related to the overblown and highly energized areas are also exaggerated, while those of the constricted areas are impaired.

The blown-up top, sitting on a small, constricted pelvis and rigid legs, gives an overall picture of a person straining upward, making himself bigger and higher than he really is. An exaggerated hunger for importance and achievement goes along with the expanded upper body. (A figure like Mussolini comes to mind.) The contracted, rigid, lower half blocks the flow of energy downward, leaving the individual ungrounded and unable at times to contain impulses. The energy flowing upward and creating ideas and activity is not properly balanced by the downward flow which gives a sense of place and connection to the world. In cases even more extreme, we have found an underlying feeling of inevitable isolation, an ill-concealed contempt for the ordinary affairs of others, and a very limited capacity for sustained feeling.

In the opposite pattern—that of the exaggerated lower half —there are other disharmonies. The tight, small, undercharged upper portion does not allow for aggressive action. The energy from below does not rise, and cannot fill the whole person. Unable to reach out or strike out, the tendency is toward passivity. The same isolation exists, but instead of impulsive action, there is inertia, an inability to take any action at all.

In these people, the way to health lies in the restoration of balance and free-flowing energy. If the rigid areas can be relaxed, the tremendous amount of held-back feeling in the overblown, overcharged, and static portions can begin to release. The relaxed areas provide a channel for expression. In the legs, these relaxed areas ground the impulsive, overblown, and isolated ego. In the chest, back, shoulders, and arms, these areas provide the means for the active expression of anger, longing, and aggression.*

Another important aspect of the total body, besides large displacements, is its asymmetries. In body-oriented approaches, we speak of a left-right split, which means a lack of normal symmetry and integration of the left and right halves of the body. An example of such a split is given in the following photos.

FIGURE 12. LEFT-RIGHT SPLIT.

* Throughout the book, we mention some therapeutic approaches that center on the body. For a comprehensive and detailed analysis, the reader may want to look into the writings of Wilhelm Reich, Alexander Lowen, William Schutz and the soon-to-be-published works of Ida Rolf and John Pierrakos.

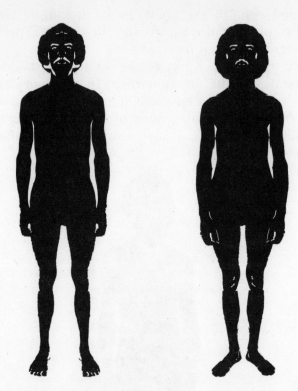

FIGURE 13. LEFT SIDE COMPOSITE.   FIGURE 14. RIGHT SIDE COMPOSITE.

The person in Figure 12 has a decided asymmetry. The left side of his body appears stronger and larger; the shoulder is higher and the whole side seems more aggressive, upright, and ready for action.

To demonstrate this even further, Figure 13 combines the left side with its own mirror image to yield a whole body composite. It shows what the subject would look like if he were perfectly symmetric with his own left side. Figure 14 does the same thing with his right side.

In these figures, it becomes completely obvious that two very distinct structures exist within this one individual. These two structures, of course, represent two strongly different trends of his personality. The left side shows a displacement upward, large shoulders, narrow hips, and an overall pattern of hardness and exaggerated masculinity. The right side reverses this, and appears softer, more curved and feminine. It is displaced slightly downward. The general structure, then, lacks the integration of these opposites. We should expect some very strong conflicts within this individual.

The body is normally asymmetric in several respects. The cerebral hemispheres of the brain perform some different functions. For example, speech appears to be predominantly a left cerebral hemisphere function. The right hemisphere is apparently less linear, more intuitive, and holistic. We associate the left side of the body (which is controlled primarily by the right side of the brain) with feelings, emotions, and the relations to the mother. The right side is associated with the father, reason, thinking, logic. Furthermore, the left is the side with which we take things in. It is receptive. The right side is outgoing, expressive, the side with which we act.*

Even with this "handedness," if the various elements are balanced and integrated, the gross symmetry of the body remains undisturbed. Where one aspect strongly dominates, say reason over feeling, or expression over receptivity, we will find left-right asymmetry.

THE FEET

Though in our culture not much attention is given to the feet, as the base upon which our entire body structure rests and as our connection with the ground, this complex network of nerves, muscles, and tendons is extremely important. We no longer walk barefoot, allowing the various nerve endings in our feet to be stimulated by every step we take. Rather, we keep our feet encased in leather tombs. This is all too reasonable a position to take when confronted with the cement of our cities, where instead of a constant foot massage, we experi-

---

* In William Schutz's book, *Here Comes Everybody*, he speculates that the left side is associated with initiating action, and the right side with following through.

ence an attack on our entire structure. Each step radiates a blow along our long bones, testing the flexibility of our knee, hip, sacroiliac, and lower back joints. Many of the ailments in these areas result from this constant trauma.

Much of how we deal with reality is expressed in the contact our feet make with the ground. If we are pushovers, our feet show this. They will be inadequate to support us. They may be too small, collapsed, or the arches rigidly held too high. In other cases, the right foot goes one way while the left foot goes another. Often, the owners of such feet show confusion as to where they are headed. Rigidity in the foot may reflect rigidity in the person. Commonly, these people have a heavy, thudding step. Theirs is an unbending approach to the real world. The rigid high arch may also reflect an inability to contact the ground and a tendency to be uprooted or ungrounded. Some people almost seem to be walking on two stilts. When the arch is collapsed and the foot caved in, this is often associated with collapse in the rest of the structure. We may speculate that in this case the individual, lacking the energy to make a firm contact with reality, utilizes his or her very weakness. Through the collapsed arch, more of the bottom of the foot touches the ground. It is an attempt, though a weak one, to experience more of life. With such a low energy state, it is really the only available method of compensation.

Some people walk by tiptoeing about, while others crash their heels into the earth. When we are unsure about another's feelings toward us, we often walk as if we are walking "on egg shells." We are being extremely careful about our contact with reality.

The feet in Figure 15 illustrate several points:

1. Starting on the left, we see feet that seem almost like a seal's flippers. As we move through the series, the feet come forward until, in the last pair, they are parallel to each other. Obviously, the feet in this last instance are able to support the weight above them much more efficiently than those in the first.

2. In the third subject, the feet are seen to be going in different directions. His whole body shows the split between

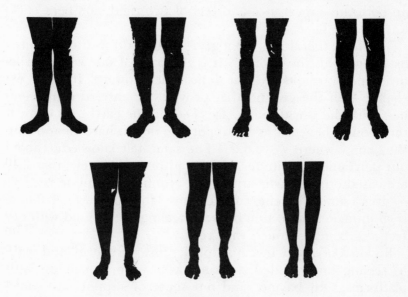

FIGURE 15. FEET.

his right and left sides. (He is the same person we discussed in the section on the left-right split, Pp. 45-47.)

3. It may be of interest to note that the feet of the first and second subjects are those of father and son. It can be understood from these photos that in some ways they have adapted to their environment with similar mechanisms.* This also brings up the question of heredity. Heredity is a significant factor in determining who we are and how we react.

4. It can be seen in the photos that with rotation of the foot forward, a simultaneous rotation of the ankle and knee joint occurs. With this rotation, the ankle and knee joint are less stressed and so better able to transmit the weight through the center of the foot. As this occurs, the entire lower extremity is able to relax and exchange energy with the earth. In fact,

* What we inherit is the substance with which we react to our environment. If we possess a low energy machine of some sort and force it to produce a high energy output, it will break down somewhere. Similarly, the body is unable to exceed the limits of its inherited constitution. In this sense, given constitutions will react to similar stresses in similar ways.

when the feet are askew, the lower extremities begin to take on both the appearance and feel of two rigid bars bent backward at the knee.

The foot, therefore, has a great deal to do with our energy exchange with the ground. It is the base of our whole structure, and transmits its own deficiencies upward. It is the exchange with the ground that provides a sense of connectedness with the world, our place. Feeling the earth, we know if the ground beneath us is supportive, or too hard or too soft. We "know where we stand." The same felt knowledge about our perceptions, attitudes, ideas, and plans—namely, that they are based on the firm, supportive ground of genuine realities —gives a similar sureness to the entire fabric of our lives. There is an intimate connection between feeling the ground with our feet and being in touch with reality.

Rigid, contracted feet disrupt the flow of energy and leave us feeling ungrounded, walking, as it were, on lifeless, stiff platforms. Our balance and our sense of support are somewhat precarious. We are uncertain of where we stand, and must live under the constant irritation of this uncertainty. Our thoughts, feelings, and actions are continuously eroded by this underlying insecurity.

Ironically, this same unsureness of the ground leads to the contraction of the feet, protecting them from unknown dangers. Likewise, we hold too firmly to our opinions and ideas, reluctant to let ourselves exchange energies with new or frightening realities.

The relaxed, flexible foot, one with good tone, makes sure, sensitive contact with the ground. It provides firm yet responsive support for the structures above, and is capable of handling whatever changes are encountered through movement.*

---

* One could translate the above sentence, using words like "ego" for "feet," and "reality" for "ground," and come up with a reasonable description of the personality of the owner of such feet. The nondefensive, flexible ego, with good strength, makes sure and sensitive contact with reality. It provides firm yet responsive support for all other psychic structures, and is capable of handling whatever changes are encountered through their interaction with each other and the world.

A FOOTNOTE TO FEET

The bottoms of our feet, like our hands, contain a great density of pores. Through these pores, toxins and waste materials are excreted from our bodies. Proper bathing and cleansing of the feet is important to maintaining good health. Proper shoes are also important. Ideally, it would be best to walk about barefooted. Unfortunately, in our culture this is impractical. Shoes, therefore, should be of good quality and of soft leather, to allow adequate flexibility in the small joints of the foot and the ankle. They should give good support to the arch and the heel and should not be elevated, as this tends to pitch the body forward, throwing the entire weight-bearing axis off and causing considerable stress on the legs and spine (especially at the level of the lower back). It is easy to imagine the pain and discomfort of the many women who walk about with their feet squeezed into shoes which are too tight, with heels that are too high.

THE ANKLE, KNEE, AND LEG

The foregoing leads us to consider the ankle, knee, and lower leg. The ankle possesses the qualities of all the joints; it is an area of motion. Specifically, adjacent segments move one in relation to the other—in this case, the foot and lower leg. When the mortice of the ankle joint is not centered over the middle of the foot, the weight of the body falls to the inside or to the outside, causing great strain on the joint.

In such a situation, the ankle may give way, leading to an experience of uncertainty when undertaking any step, particularly a new one.

In the individual who segments his or her body flow by building up rigidity in the joints, the ankle is almost invariably involved. In the extreme, the person's movement seems like that of a toy soldier. As a general rule, the joints are an important key to the way we move and, therefore, how we look to others. In this sense, they reflect style and personality. The movements of the drill sergeant, the vamp, the drunkard, or a political candidate tell us a great deal about these people. In addition, we know intuitively be means of our own structure what their statement of themselves is.

In the series of legs shown below, the significant factor to be noted is the degree of locking in the knee joint. The term "locking" is quite accurate; the knees are pushed back and tightly braced. This is particularly true of the last subject on the right.

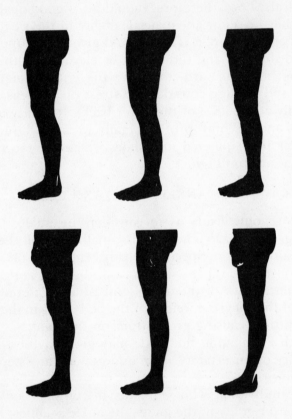

FIGURE 16. THE KNEE AND LEGS.

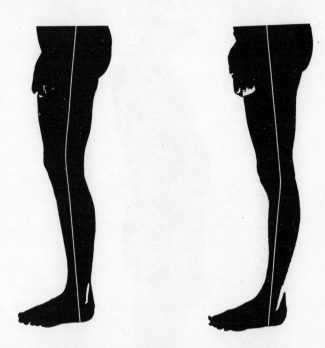

FIGURE 17. ANGLE OF THE KNEE JOINT.

The mechanism behind locking of the knee varies with the basic emotional and structural pattern of each individual. In some of these mechanisms, the person uses the locked knees to:

1. Keep from being suppressed or subjugated—"I *will* stand up! I *won't* give in!"

2. Allow him to stand his ground no matter what—"I won't be put down!"

3. Maintain a hold on reality—"I must hold myself together."

4. Keep the already collapsing structure from falling—"This is all I can do to hold myself up." "You should help me!"

FIGURE 18. WILLFUL HOLDING.

In Figure 18, it is possible to sense how this person is willfully holding on. His body is bowed backward, his feet are grasping the ground, and his knees are locked. One can imagine a force acting on him which he opposes. It is as if he is making a constant effort to stand his ground or move forward.

Independent of the mechanism, in many people, the energy in the legs is being tightly contained, and the flow between the ground and the organism is partly or severely limited. Lacking flexibility in the ankle and knee joints, the entire lower extremity moves as a single block. (With flexibility, a cyclic receiving and giving to the earth occurs with each step. Energy flows and the organism moves fluidly.) Below the locked knee, the tension in the leg is quite marked, and there is an associated rigidity in the feet. When people with rigid legs are asked to bend their toes forward or back, they can barely do so, since movement in the toes is controlled by the muscles in the lower leg (muscles that lie along the shin bone.)

Another association with the knee, in terms of emotion, is the well known fact that, with great fear, our knees begin to shake. In ancient Chinese medicine, the knees were felt to be related to the kidneys. The kidneys in turn were related to both the water element and the sexual organs. Water in this system relates to fear and correlates with the above. Our sexuality is related to our deepest vitality: forcing a man to his knees has always been experienced as humiliating. Going down on one's knees before a king or holy person or sacred image is a traditional sign of submission. We may reasonably speculate that locking the knees is also related. "I will not bend to your will." "I will not beg you."

Let us now examine the section above the knees, the thigh, and buttocks.*

---

* The buttocks make up the muscle mass which goes from the pelvis to the thigh posterially.

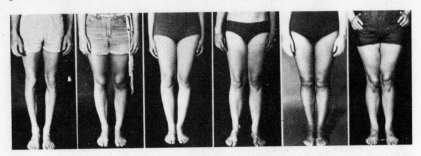

FIGURE 19. THE FRONT OF THE THIGHS.

In this series, we can begin by noting the space between the thighs. In the first subject, the knees are quite separate and the space terminates at a high peak in the midline. Visualizing this as a high arch (gothic arch) is a useful image. In the second and third subjects, this space decreases. The fourth is the closest to a normal alignment. In the next subject, the knees are close together and tension is noted in the thighs, including the areas of the groin. In the last, no space is seen between the thighs.

Where we have the high peak (gothic arch), the tension is present at the center point (in the area known as the perineum). It is here that the holding exists. The organs involved in this area are the genitals, anus, and rectum, with the greatest tightening immediately in front of the anus. With spasm and energy blocking here, the sexual function is prevented from finding full expression. In the last two subjects, the tension in the thighs draws the thighs inward, squeezing the genitals. Here again the genital expression is limited. The normal situation allows the area of the anus and genitals full flow, the thighs not being too close nor sprung apart. The experience in the two situations is quite different. In one, the thighs actually are squeezed together protectively. (This is normally not a conscious act, but an ongoing, unconscious statement.) In the other, the inner thigh muscles are shortened and sucked into the center point.

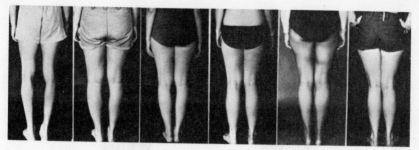

FIGURE 20. THE BACK OF THE THIGHS.

If we view these same subjects from the back, in the first two we can clearly sense the effect of the gothic arch jamming into the area in front of the anus.

Subjects three and four are transitional in that there is far less upward jamming. In the last two, we begin to note that the buttocks are squeezed together. A clue to recognizing this can be found by observing where the fold of the buttocks lies.

If the reader stands for a moment and squeezes his buttocks tightly, simultaneously squeezing his thighs together, he will have a good sense of the last three subjects. The amount of tension is considerable, and the effectiveness of the contractors in blocking the flow of vitality is apparent. Again, remember, the contraction is present even though the subject is unaware of it. It is the "unconscious" manifesting in structure.

The chronic contraction of the large muscles of the buttocks consumes a large amount of energy. They are highly charged and often relatively light pressure (particularly on the inside of the thighs) evokes strong emotional responses. On other occasions, the surface area over these muscle groups appears to be soft and flabby, belying the underlying tension. With deeper pressure, they are found to be still more tense and tender. In this case, the use of massage in depth will allow these held charges to release. Often, the released emotions are quite intense, frequently being those of rage and anger. The muscles of the thigh and buttocks fix onto the pelvis and are

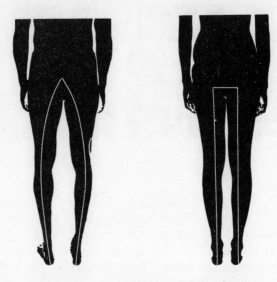

FIGURE 21. GOTHIC ARCH AND NORMAL.

very important in determining both its position and mobility. When chronically contracted, they effectively immobilize the pelvis, again reducing the organism's ability for sexual release.

THE PELVIS

It should be clear by now that tension in one area reflects back onto many others. In a sense, even a single tension is never really isolated because it includes compensations. Perhaps the two most contained expressive movements in the body are those of the pelvis and the breath. The more breath we have clearly indicates the degree of our involvement with life. Modern, industrial man cuts off his breath. This is the most direct way to contain rising influences and feelings, as every breath-holding child knows.

The unblocked pelvis is able to swing freely forward and backward when walking to allow an easy to-and-fro motion, in contrast to a side-to-side wiggle. In this latter case, the heightened mobility, possibly suggesting greater sexual freedom, may only indicate an increase in local energy charge. This charge, however, occurs without integration with the rest of the body and is more often associated with an underlying hysteria than with increased sexual appetite or sophistication. Hysterics tend to dramatize with overactivity. In some other situations, the pelvis does not really move. It is carried about in one solid block, usually locked in a forward position (tucked under) or retracted backward, cocked, as if ready to fire. This firing, owing to the lack of mobility, is always poorly accomplished. The following sequences demonstrate various positions of the pelvis.

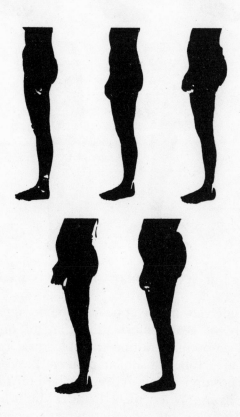

FIGURE 22. POSITIONS OF THE PELVIS, FEMALE.

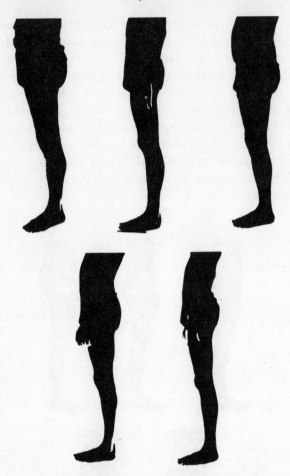

FIGURE 23. POSITIONS OF THE PELVIS, MALE.

From this side view, there appears to be no significant difference between the sexes, illustrating that in some aspects of physical and emotional trauma the structural compensations in the pelvic area follow a similar pattern. Basically, the pelvis assumes the position which expresses the individual's set attitude toward living. When tucked under, tight buttocks allow for only a dribbling out of emotion and feeling. The individual is unable to allow the pelvis to swing back and gather strength for the forward thrust associated with full emotional discharge. Emotions can at best be squeezed out. The opposite extreme is one in which the pelvis is retracted, or held back. The individual is unable to release. A great charge has been gathered in the pelvic region but is unable to swing forward. We may use the analogy of a gun, cocked and ready to fire. Perhaps this very clear parallel can be taken a step further. These individuals may be viewed as being afraid to let go, since the high charge they are carrying may be sensed as explosive. This anxiety is often associated with the question, "What will happen if I let go?" When such an individual is asked to describe what might happen, the reply is often "I don't really know." The sense is one of an unknown force. Further examination often reveals fears of sexual promiscuity or violence toward others.

Figure 24 shows two people with very different pelvic structures.

The first person's pelvis is close to normal—that is, squarely under the trunk in a horizontal position. In the other person, the pelvis is retracted, held back. When considered in terms of size alone, the pelvis may be too small or too large, relative to the rest of the body. This usually follows the pattern of top-bottom displacements we have already discussed. In general, a tight, contracted, and small pelvis is associated with either immaturity and a lack of development of feelings of sexuality and instinctual drive, or with a strong containment of these instinctual feelings.

At the other extreme (particularly in women), we may find the pelvis to be excessively wide. Here the sacral bone is pushed forward, spreading the pelvis. Women possessing this structure, in our experience, tend to be matronly, in a sense, hyperfemale, with a great deal of deep warmth. For them, the pelvis is a repository of tenderness and nourishment. Associated with this increased width of the pelvis, are piled up, soft, pasty hips, buttocks, and thighs. We conceive of this flab-

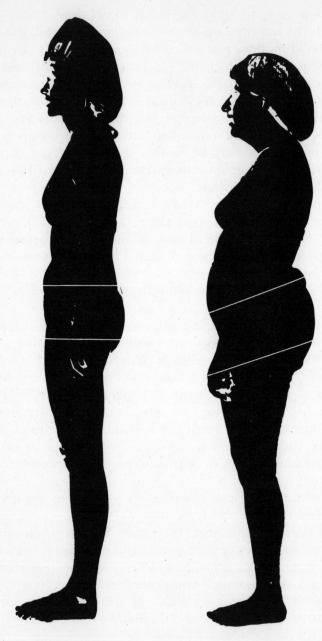

FIGURE 24. THE ANGLE OF THE PELVIS.

biness as an exaggeration of the passive-receptive position and have found many women who indeed fulfill this image. They are almost archetypal figures. (The painter Rubens often portrayed such women.) Others, while possessing this fundamental trait, struggle with this passivity and attempt to overcome their flabbiness with action, whereby a nice harmony between receptivity and action results. In others, such attempts most often end in feelings of hurt and frustration.

RESPIRATION, THREE CENTERS, AND THE BELLY

In our culture, the idealized body form has a narrow waistline and a flat belly. Vast sums of money, much time and great effort are expended to reduce abdominal girth. Not so many years ago, women's corsets held their bellies in unremitting strangleholds. Even today, the elastic girdle continues to do a landslide business. BELLY IN–CHEST OUT is the script we attempt to live by.

FIGURE 25. NORMAL RANGE OF RESPIRATION.

Normally, the belly wall expands with each breath. To avoid this and still remain within the limits of our culturally ordained script, we consciously or unconsciously keep the chest inflated and the belly muscles tightly drawn in. Let us compare the normal free flow of breath with the "held" type of breathing.

In Figure 25, the shaded area show the normal range of respiration. In held respiration, the diaphragm hardly moves at all and the belly stays "sucked in." The chest may also be under- or overinflated, and there is very little movement. Contrast this with the latitude of normal respiration where the flow of breath carries into the abdomen, allowing the belly wall to relax outwardly. The diaphragm is free and mobile. In primitive societies, in marked contrast to ours, breathing is seen to be relaxed and normal. In fact, the more primitive the society, the greater the ease with which the belly wall moves. In our world, the most natural state is found in very young children.

What then, in our culture, is the underlying reason for the emphasis on shallow breathing and a tightly contracted belly? Is it really a preference based on vanity? Or, as in other areas of holding, does it serve to block off feelings? We begin to discover answers by observing that the belly is the most exposed and least protected area of our body. It is soft and tender and vulnerable. Within it, vital organs are contained, including our intestines, or "guts." In our culture, the individual is rarely supposed to express his "gut" feelings. Neither is a time or place provided for such expression, nor is it normally encouraged to continue to a resolution when it does appear. On the contrary, we are told to "calm down" and "pull ourselves together." Fear of emotional expression pervades our entire way of life. We are prone to call people "crazy" when they give vent to outrage, despair, or hopelessness. The natural course of such expression, the "cleaning out" of this backlog of human misery is more often arrested than given the benefit of support and understanding. In such a circumstance, the recovery of one's mental balance becomes difficult indeed.

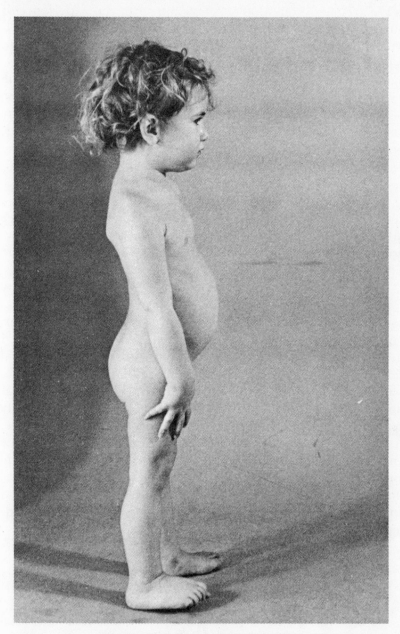

FIGURE 26. BELLY OF YOUNG CHILD.

We are conditioned not to allow ourselves the experience of our own gut emotions. Many people with whom we work, when encouraged to breath softly into their bellies, begin to release deep feelings. As the breath flows in, the contained emotions flow out. Often, trembling in the lower half of the body takes place. The pelvis may begin to jerk spasmodically, or rock, and the legs vibrate. This is usually accompanied by anxiety or panic at first, marked by a holding of the breath. If, however, the person is encouraged to breathe and continue his experience, in time the anxiety and panic is replaced by an ecstatic feeling of warmth and wholeness.*

In schools of Eastern knowledge, contact is made with a center in the belly located in the midline, several finger breadths below the navel. This center has been called by several different names: *Hara* in Japanese Zen, *Tan-Tien* in the Chinese system of Tai Chi, and *Kath* in Sufi schools.

In these schools, this center is felt to be a center of vitality related to our basic instinctual drives. When the student has come into full contact with the energy and force of this center, he enters a new level of consciousness. He or she is grounded in this center, and it serves as the spiritual and emotional center of gravity. In our Western, more body-oriented therapies (such as rolfing or bioenergetics), as the blocks to respiration in the chest and diaphragm are released and this belly center is contacted, the individual experiences a deep change. As we have already noted, the initial reaction is often one of anxiety. With courage and lots of work, this can be overcome and a new experience occurs. The lungs take in more breath, the genitals become more alive, the legs receive the breath's energy, and the guts loosen from the panic of constantly holding on.

This belly center may be viewed as having a "mind" of its

* It is essential for the therapist to remember that this passage from fear to ecstacy or pleasure can occur only when the therapist experiences within himself his own gut feelings. This allows him to approach the patient with the deepest compassion. Knowledge without compassion leads to a sterile cerebral therapy where the "gut" emotions remain locked behind a wall of muscular spasm. Compassion, on the other hand, allows the therapist to serve as a vehicle for the liberation of his patient.

own. This "mind" includes the consciousness of our basic instinctual drives, our sense of hunger and satiation, our sexual drives, our "gut" awareness of others and of our environment. In contrast to this belly mind, our head mind or thinking mind has ideas about what, when and how to satisfy our basic instinctual needs. Rather than our guts telling us when and what to eat, our head does. Between our belly mind and our head mind lies our chest or heart mind. This is the seat of our emotions. Our feelings of envy, jealousy, greed, vanity, and sloth—as well as love, generosity, inner peace, and emotional balance—are centered here. Often, the conflict which arises between the belly mind and the head mind is expressed by the emotions of the heart and chest. Our belly mind might feel sexually fulfilled once weekly, whereas our head mind pushes us to have more. Our chest mind, caught in between, experiences this as an unbalanced desire. Sex, lust, jealousy, envy, gluttony all begin to find expression as deep, heartfelt emotions. Chatter and static fill the organism. The heart begins to pound. The respiration may become shallow and rapid, or suspended. Anger and irritability rise at the least provocation. The belly mind, in the face of this raging turmoil, attempts to protect itself by contracting the diaphragm and the belly muscles, thereby cutting down the emotions flooding it. As these unbalanced emotional demands persist, the belly and the guts contract more and more. This may sometimes result in the development of disease states such as colitis, ulcers, or even cancer.

In summary, we may say that the belly mind contains a fundamental, instinctual consciousness. By means of this consciousness, it attempts to maintain its own integrity, cutting off those energy impulses from outside itself (chest and head) which are not in harmony with its own needs. True balance in the body occurs when the three centers, head, chest and belly, are functioning harmoniously. As we are all aware, in our contemporary culture, the mind is very highly trained and placed in control of the lower, instinctual belly mind, resulting in often confused, unbalanced emotional responses. In the reverse situation, we act from our instincts, or basic energy drives. Our heads follow this energy, integrating internal and external information so as to attain body satisfaction. Our emotions in this case reflect our body's true energy needs.

There are three areas in which energy streaming can be cut off: the throat, the diaphragm; and, the lower abdomen. In the throat area, tightening occurs every time we are asked by our head center to say something emotionally difficult or phony, i.e., not in tune with our internal instinctual life. The diaphragm tightens whenever our gut feelings are supplanted by our head's ideas. The area across the lower abdomen tightens, freezing the pelvis and cutting off genital feelings whenever our head dictates when, how, and where to have sex without harmonizing the dictates of our belly and heart centers.

It seems to the authors that the pervasive miseries of contemporary culture, our terrible isolation and alienation, our cold destructiveness, are due in large part to a lack of contact with our own instinctual gut life. As long as we continue to overeducate with mind structures, our vitality will suffer. As long as we remain on this massive *Head Trip,* no real solutions can evolve. Only by recovering the knowledge within, and adjusting modern life to its requirements, may great change occur.

The importance of the right relations among these three centers was known to the ancients. A card from the tarot deck may serve as an example.*

In the card called The Hanged Man, the man is upside down. His belly center is on top, his head center on the bottom. About his head is a luminous halo. This card states that the passage to enlightenment (which means proper balance of the centers) requires dominance of the belly over the head. Of course, being upside down can be very uncomfortable. It takes time for the shifting of dominance from head to belly center to occur. This painful period is also symbolized by the hanged

---

* Tarot cards have been used by mystical societies for thousands of years, possibly making their first appearance in Egypt. They describe man's passage from spiritual sleep to an awakened state.

man's position. In our clinical experience, the patient may indeed go through a phase of feeling "upside down," as the anxiety of surrendering control of the rational mind slowly dissipates. Our heads have been in control too long and will not let go without a deep struggle. We believe it is this overemphasis on control by the head which is the real, underlying reason for our almost adament preference for that tight, narrow, and trim waistline.

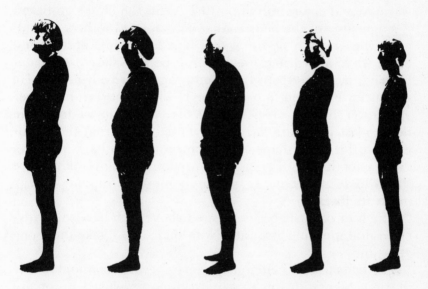

FIGURE 27. THE BELLY (MALE SERIES).

Since the tight, constricted waistline interferes with breathing and belies a fear of full emotional expression, deviations from this ill-conceived ideal reveal other psychological traits. The shape of the belly can be quite variable, as can be seen in Figure 27, a series of male subjects. Of all the myriad, possible shapes, we have delineated four basic shapes.* These are: enlargement of the upper half; enlargement of the lower half; overall enlargement; and flat.

* This delineation, as well as some of the clinical observations which follow, are taken in part from the work of George Groddeck. See *The Book of the It*.

Since very little has been done in this particular area, only very general remarks can be made and these only tentatively. However, one firm impression does arise, and it is that the relaxed, evenly toned, not obese belly—one that allows the respiratory wave throughout—is associated with full vitality and openness to feeling. It is this belly one finds in the innocent, the very young, and the untroubled primitive.

Of all the factors which seem important in assessing the meaning of various belly shapes, such as size, motility, tone, warmth, and color, overall posture seems the most significant. A single area of the body cannot be evaluated without considering the whole. A person with a sway back and a protuberant belly, while appearing open, cannot be considered to be unblocked, moving breath freely into the lower belly center. He will surely be found to have some blocking between his upper and lower halves somewhere in his back. Subject four, the next-to-last in Figure 27, when viewed only from the standpoint of his belly shape, might be thought to have an open, even, smooth flow of energy. Inspection of his overall posture shows he is quite bowed, and surely must be trapping energy in his midback.

The four basic shapes delineated above each have some psychological meaning associated with them. We'll take them one at a time.

The belly which is enlarged in its upper half is found mostly in men. Men with this shape are often employed in, or are used to, heavy physical labor. They are generally rugged and quite masculine in appearance. (See subject one in Figure 27.) If the reader compares his shape with that of the child in Figure 26, the same enlargement above the belly button will be seen. We may speculate that these rugged men still carry about with them the child they once were. Perhaps as Groddeck says, in every man, there is ultimately a little boy. This shape can also be part of a general thrusting upward found in people with large segmental displacements.

When the enlargement occurs in the lower half of the belly, we have our second basic type. If it is extreme, it indicates that the energy flow into the lower pelvis and legs is strongly blocked. In others, where the lower belly is full but not obese, warm to the touch and receptive to the flow of breath from above, the individual may be expected to have good contact with his vital lower belly center. It has been speculated by Groddeck that, when extreme enlargement occurs in the female, it may indicate a deep, unconscious desire for pregnancy. (Our own clinical experience with this type is too limited to allow comment.)

None of the subjects in the male series fits this belly type, but in Figure 28, a series of female subjects, all show it. Subject three in this series, while showing a lower belly exaggeration, begins to approximate the third basic type, an overall enlargement.

While some degree of obesity is present in this subject, this form may seem to be an idealized third type. There is an important distinction here, however. Overweight, in our experience, with a heavily padded abdomen, is most often asso-

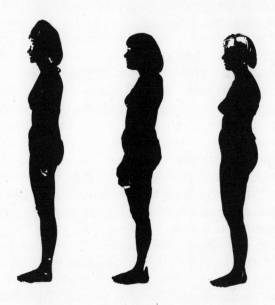

FIGURE 28 THE BELLY (FEMALE SERIES).

ciated with a lack of contact with the belly center. It is probable that this very lack of contact makes for the individual who constantly overeats to fill up. Where the enlargement occurs throughout, without obesity, the shape is compatible with good energy streaming into the lower belly center, and an index of good health and vitality. We have already discussed the last basic type, the flattened, contracted belly. The' last subject in Figure 27 is a good example.

THE EARLY BOND

The umbilicus, or belly button, divides the belly into upper and lower halves. This structure serves as a constant reminder of our origin. We are made of flesh and blood and entered this world tied to our mothers, receiving nourishment through the umbilical cord. At times, this nourishment might have been less than it should have. The mother who, for whatever reasons, experienced negative emotions toward her unborn child, even as it grew within her, might very well have cut off some of the life energy she had available for this purpose. Sometimes poor physical health or emotional distress unrelated directly to the child can result in decreased amounts of energy input to the growing fetus. We carry this inheritance within us as a quantum of life energy. Hence, on a more spiritual level, the umbilicus reminds us that we are all children of our physical mothers, but that, for a short time, we are physically free to journey through life experiencing whatever growth and fulfillment we are capable of. It is also a reminder of our deep ties to our spiritual or cosmic mother, the Creative All in which each of us is immersed. We are living right now in that womb which is this universe and all the energy contained within it.

Our relation to our cosmic mother closely parallels that to our physical mother. In that first connection is the seed of all later expectations and fulfillments. If, in the uterus and early extrauterine environment, we were poorly nourished, we will find it emotionally difficult to receive any energies. Somewhere within us the experience of not receiving full warmth and love is imprinted. We will be untrusting, unable to open ourselves to the available nourishment around us. The authors have seen a number of individuals who, in their physical makeup, reflect this lack of direct, umbilical nourishment.

Even after birth, the early relation to the mother has the quality of an umbilical tie. It takes the newborn infant a considerable time to differentiate itself from its mother. Until it does, mother and infant may almost be considered a single energy field. Lack of a satisfying exchange within this field can be seen in several ways. One primary way involves all the characteristics that result from having low levels of energy. Thinness of the body, weakness, underdevelopment, dependency, a collapsed physical structure, and an inevitable clinging to others for support are some of the results. All of these follow directly from lack of nourishment and the low levels of energy such a lack entails.

THE INTERNAL ORGANS

Certainly other physical qualities also tell us about this early bond. These have to do with the relation between organ functions and personality. The early Greeks, for instance, recognized four temperamental dispositions: sanguine, melancholic, phlegmatic, and choleric, each respectively dealing with the blood, low spirits (energy), phlegm, and bile. A much more sophisticated system can be found in Chinese medicine. Reaching back to antiquity, we learn that all the different body organs are associated with different mental and emotional qualities. When an organ is not functioning properly, a change in skin color, particularly around the eyes, temple, and mouth takes place. The color, or shade of color, varies with each of the organs. Though subtle, they can be recognized.

In the case of an individual whose early tie with the mother was disturbed so as to produce a lack of adequate emotional and physical nourishment, the color most often noticed is pink. This is the color of the small intestine which, before birth, may be viewed as extending into the umbilical cord. In the adult, it is the major organ of absorption. Through it, we receive nourishment from the foods we ingest. We are unable to survive without it. In one individual with a pink hue, the profile of the belly wall showed a definite protrusion at the level of the umbilicus. Perhaps the triad of (1) psychologically experiencing the mother as cold and unnourishing, (2) a pink hue, and (3) a protruberant abdomen at the level of the umbilicus, can help to identify these individuals. Another helpful hint is that most of these "pink" individuals are overweight, perhaps because they seek the nourishment they never had. Clinically, the relation of pink hue to cord dysfunction has never been established to the authors' satisfaction. The experience of some 3,000-plus years of Chinese medicine, however, tends to lend additional support to this relationship.

Table 1, which follows, lists bodily organs, their related psychological functions, and the subtle skin colors which indicate dysfunction. (For further information see *The Yellow Emperor's Classic of Internal Medicine*, University of California Press, 1972.

TABLE 1

ORGANS, PROJECTED COLORS, AND PSYCHOLOGICAL FUNCTIONS

Organs	Color of Dysfunction	Psychological Functions
Fire Organs (Red Group)		
	Red, with slightly Bluish tint	Contentment, joy, impulse controls, emotional life.
Small Intestine	Pink	Affection, connection with physical and spiritual mother, absorption, separation of pure from impure.
Triple Warmer	Orange-Red	Strength, vibrancy, affection in relations with spouse, children, and friends.

| Circulation Sex (sexual function and hormonal regulation of blood) | Purple-Red | Sexual energy, emotional vitality, protector of the heart both physically and emotionally. |

Earth Organs (Yellow group)

| Stomach | Deep Yellow | Asks for affection, sympathy. |
| Spleen | Orange | Movement of energy, enthusiasm as opposed to apathy. |

Metal Organs (White group)

| Lungs | Brilliant, pure white | Controls energy and breath, inspiration (emotional fatigue and claustrophobia). |
| Colon | Abalone, shell white | Eliminates psychological wastes, acts rhythmically when we are in the "flow." |

Water Organs (Blue group)

| Bladder | Deep blue | Keeps us from drowning in our emotions. |
| Kidney | Pastel, light Blue | Stores energy, cleans psyche as well as blood, will power. |

Wood Organs (Green group)

| Liver | Very dark Green | Clarity of vision, sense of security (or insecurity), ability to organize appropriate plans (anger and jealousy). |
| Gall Bladder | Pale Chartreuse | Ability to make independent decisions, clarity of thought. |

In order to further clarify the above, we are including the following brief synthesis of the relationship of the organs to physical and emotional functions, as seen in Chinese traditional medicine.* Man contains within himself the five basic elements of which the universe is composed. These are Fire, Earth, Metal, Water, and Wood. Their respective colors are red, yellow, white, blue, and green. When placed in a circle, as the Chinese do when viewing these elements, their interrelationships can be readily seen.

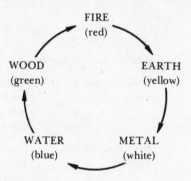

Water nourishes Wood and Wood feeds Fire and from the Fire comes the ashes producing Earth and its contained minerals (Metal). Each of these elements in turn contains several organs or metabolic control systems. As we have already mentioned, associated with each is a color shade of its elemental color. Hence the small intestine, which in this system is a fire organ (red element), has a pink color. The Chinese viewed man as a unitary being whose physical and emotional derangements reflect different aspects of imbalance in his vital life energy.

In the case of the nervous system, there is little clinical experience which goes beyond the observation that disruptions in function are associated with the colors violet and lavender.

* We are indebted to Professor Jack Worsley of the Chinese College of Acupuncture, United Kingdom, for his enlightened teachings which brought this material to life for us.

The associated psychological functions are, as would be expected, related to emotional balance and movement.

With the use of the above system, it is possible to make an analysis of an individual's organ function and to correlate this with aspects of his emotional and psychological state. In a sense, this system allows us to look directly into the core of a person. For example, an individual experiencing a crisis was referred to one of us. She came in completely overwrought and fatigued; she had been depressed for three weeks. Feelings of despair would overwhelm her suddenly, and she couldn't stop crying. Upon looking at the colors around her eyes and temples, a pale, pastel blue could be seen. (Her face overall had relatively healthy color.) She had lost about ten pounds and appeared very thin around the waist and lower abdomen. The pale blue color suggested a urinary problem. Her posture, thighs turned in, pubic drawn back and belly tightly contracted, strongly suggested someone who needed to urinate and was holding back. She very quickly identified with the posture, remembering the problem of bed-wetting as the most trying ordeal of her childhood. A few simple body manipulations and a full, expressive release of hatred were enough to return her to some feeling of normalcy. In this individual, the connections among posture, organ dysfunction, subtle face color, and psychological functioning seemed very clear.

THE DIAPHRAGM AND CHEST

The diaphragm is a part of the core structure. It is a muscular and tendinous sheath stretching across the body from the backbone to the ribs in front. It separates the heart and lungs above from the abdominal cavity below, and is of vital importance in regulating energy flow in the body. When it is not moving freely, the breath wave is arrested at the level of the lower ribs. (See Figure 25, page 63.) In our experience, significant progress in therapy begins when the diaphragm begins to move freely. Before the blocks in the diaphragm are released, the individual is unable to vitalize his lower belly center and give in to the nonrational, instinctual life within him. The term "schizophrenic" means split mind or split heart. It might just as easily mean a split at the diaphragm. Almost all severely disturbed patients, to our knowledge, show severe blocking at the diaphragm. Certainly, the schizophrenic, who is sometimes capable of doing great physical harm to himself with no apparent pain, is split off from his lower belly, his instinctual center. In our society, where an ever increasing degree of personal and interpersonal alienation is endemic, it is not surprising that a block in the diaphragm is the most common clinical finding. The person with emotional problems, almost without exception, has dysfunction in his breathing.

Below the diaphragm is the great cavity of the abdomen, containing the organs of metabolism and sex. The organs of digestion, also contained in the abdomen, particularly the small intestine, provide the fuel and fire needed for life.*

Above the diaphragm is the chest cavity, containing the two great pumps of the body: the heart and the lungs. The lungs take in air, feeding the fire organs, and thereby maintaining a living flame within us. The meeting of air and fire takes place just over the diaphragm, lighting the base of the heart, (the flame of a candle within us) and giving it the energy of life. When our hearts are open and unrestrained, the energy of that flame moves throughout our entire being. Our skin is warm, our voices joyous, our eyes sparkle, our temperaments both mellow and nourishing. So, within our chest is the meeting place of fire and air. The great pumping of the lungs reaches out and sucks energy from the air, then transmits it to the heart which, in great bursts, propels it throughout the body.

Hence, an understanding of the movement of the chest and lungs is essential to sensing the totality of the life force or vitality in an individual. It is within the chest that we sense the energy of our impulses and passions. Note the intensity of the

---

* The sex organs are also fire organs, containing the heat and fire necessary for the creation of new life. Through them, the greatest alchemy occurs; a whole, living being is created by a process of biological fusion, in which the potential for consciousness is included. This is the evolutionary jump of humanity.

expression, "My heart throbs for you." The emotions of inspiration sit within the chest. Of course, this includes our visions, our connections with the energies around us, our intuitions, and our spirituality. To "inspire" means to take in the Spirit.

The chest may vary in the amount of air it takes in at each breath, a result of the amount of energy being expended, and the amount of oxygen needed. A resting breath should provide the individual with enough oxygen to keep his tissues well supplied. It should also provide enough of a breath to provide the organism with sufficient quantities of the subtle prana energy. However, what we have learned from our daily working with people is that insufficient breath, coupled with spasm of the diaphragm, is an almost universal condition. It is as if we dare not live too much. We keep ourselves half dead. When we exert ourselves, we find that we cannot sustain any efficient input of air for very long. (Which of us, at age 30 and without training, can run even a third of a mile without collapsing?) The candle within us flickers very quickly. The authors are convinced of this basic concept of the ancient discipline of Hatha Yoga: many diseases would disappear if man were to breathe fully of air (and prana).

It is a clinical fact that when people in therapy begin to breathe fully, they increase their available energy quite rapidly. In fact, just by encouraging breath alone we often find that many people come in contact with repressed emotions and feelings. The augmenting of the life flame within the chest begins to mobilize the contained emotions of the heart and lungs. One of the emotions that arises frequently is that of longing, longing for love and affection. Deep, heartfelt sobbing occurs, sometimes freeing the withheld sobs of a lifetime. The child whose sobs were cut off long ago often reappears. These are moments of great emotional impact for both patient and therapist. (If the reader will take a moment to sigh deeply, he may experience briefly the sense of melancholy and longing found in the chest.)

The chest ideally has an oval shape. When viewed in cross-section, its side-to-side diameter is greater than its front-to-back diameter. With breath, the chest expands in both these

directions. It also elongates along its vertical axis as the diaphragm descends. The ribs, in turn, open and close much like
a Venetian blind. While all of the above is happening, the
spinal column also elongates and straightens the lower lumbar
spine, while rocking the pelvis forward. This sets up a wave of
motion along the whole spinal cord, bringing with each breath
a pulsation into the central nervous system (spinal cord and
brain). In fact, spinal fluid is actually pulsed along the central
canals of these structures.

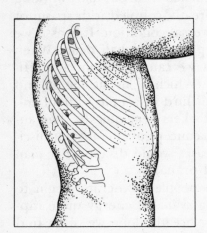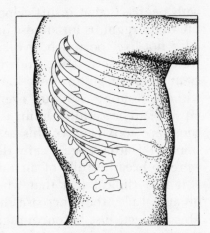

FIGURE 29. MOVEMENT OF RIBS DURING BREATHING.

Here, then, we see the interrelationships between breath
energy and energy in the nervous system. Each breath brings
an impulse, charging the nervous tissues. When an organism
is breathing fully and charging itself, the eyes, which are a
direct mirror to the nervous system, become bright and shining.* Unfortunately, for the many reasons mentioned throughout this book, the chest does not expand adequately in most

* The eyes are actually an extension of the nervous tissue of the brain.

people. Not only is the volume of breath decreased, but the mode of distribution of breath is also hampered. For example, if the breath wave does not include an appropriate descent of the diaphragm with elongation of the spine and rocking of the pelvis, the nervous tissue will remain undercharged. This is true even if the person breathes very rapidly. The importance of the form of the breath wave should not be underestimated. In spite of many variations and minor details, three basic types of chest structure can be noted: overexpanded; collapsed; and asymmetric.

The overexpanded chest shows very minimal collapse during expiration. It is held in a tight, inflated position forming, as it were, a wall of protective, hard muscles around the heart and lungs. This wall, involving the muscles of the back as well as those of the rib cage and the front of the chest, completely encircles those organs. The heart and heart feelings are kept locked up behind this wall.

In individuals with overexpanded chests, we find a fear of taking in energy from the outside, which involves the energy of relationships. The fear is one of softening, letting down one's guard, letting up one's pace. These people "stay within"; they stay within the bounds of rules and schedules (many of which they have devised for themselves); they stay within a rigid system of attitudes about how people should act; they stay within a rational, logical, intellectual framework, which almost always fails to comprehend the emotional and intuitive aspects of their interactions. There is a determined pride about these people, and a strong emphasis upon performance and success.

The more held and expanded the chest, the less fluid the personality. The most striking observation about these individuals is their inability to exhale, to release the air contained within. The chest resists collapsing. These individuals do not let go very easily. When they do, either through encouragement or the application of direct force, deep sobbing can result. A longing to be free, not too surprisingly, is often felt.

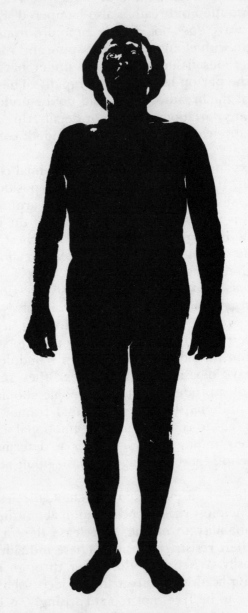

FIGURE 30. THE OVEREXPANDED CHEST.

The collapsed chest is most closely associated with a basic lack of emotional vitality. The amount of breath taken in is not adequate enough to spark full feeling. Skin color is generally pale and the eyes appear clouded over, or dull. The relationships of such individuals lack vibrancy. The low level of breathing leads inevitably to a low level of energy.

In physical appearance, and often in their early emotional history, it is as if these people have received a blow into the center of their chests. They have been deeply hurt. Their hearts are sunken back and closed off. The general feeling they project is one of tiredness, hurt and need for support. Depression, emotionally and structurally, in the center of the chest marks this type of individual, and deep breathing brings this individual into contact with his hurt. By limiting breath, facing this pain is avoided. Once he or she begins breathing more fully by going through the experiences of feeling and expression, and then integrate this deep hurt and the pain and fear of pain that goes with it, their energy level changes. They become more alive.

The third group, the asymmetric type, is really composed of several subgroups. (An example can be seen in Figure 12, page 45.) The asymmetry between right and left sides, as well as top and bottom, is clearly seen. In people with marked asymmetry, the form and depth of respiration is impaired. Most frequently, the psychological features of the collapsed type of chest are found. The asymmetry of the chest generally translates itself into a total body asymmetry. In their emotional makeup, these individuals show a lack of balance, and splits in their attitudes. In general, a twist in the body is associated with psychological and emotional twists.

As we've said, we have never seen a distrubed person who did not have some abnormality in his respiration. Perhaps this is related to the fact that the chest center is the seat of the emotions, where the energy of the air as oxygen (and prana) fuses with our material being, giving it the spark necessary for healthy functioning. In terms of Western physiology, the syndromes associated with reduced ventilation, such as those of

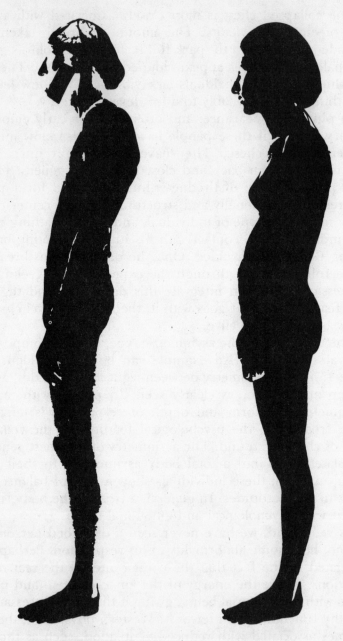

FIGURE 31. THE COLLAPSED CHEST.

heart and lung failure, are also commonly associated with fatigue and emotional instability. The so-called "neurasthenic" personality displays respiratory difficulties, without doubt. He or she is described as having a slender, underweight, lanky body, tachycardia (fast heartbeat), low blood pressure, low temperature, and basal metabolic rate. He is further described as having intermittent attacks of chest pain and rapid, shallow breathing. In some cases, these so-called "asthenic neurotics" are considered to represent cases of early schizophrenia.* The above clinical description comes quite close to describing many of the features we encounter in type two chests. These latter chests are also frequently associated with oral or low energy character type we have already described.

THE SHOULDERS

Across the top of the chest, in fact, riding on it like a yoke on oxen, sit the shoulders. They are part of the doing or extrinsic system, and should be free to move without constraint by the muscles joining them to the chest. It is only with free, unimpeded motion of the shoulders that we are able to take full advantage of our arms and hands. It is our hands, with the opposing thumb that ultimately allows us to develop creative expressions denied other species. Reaching out, grasping, the use of tools and hand-held weapons are intimately connected to the functions of the shoulders.

The shoulders are traditionally associated with work ("putting one's shoulder to the wheel"), responsibility ("shouldering" one's burdens) and, in general, doing. The strength of a man is seen in broad, muscular shoulders. One's capacity to aggress, to reach out and take what one wants, to strike out in anger or to remove obstacles—all involve the shoulders. When fatigued, overburdened, or hopelessly blocked, the shoulders show it.

Several structural deviations from the healthy can be ob-

---

* From *Current Diagnosis and Treatment* edited by Marcus A. Krupp, M.D., and Milton J. Chatton.

served. In these cases, the shoulders may be pulled back, forward, up to high, they may be too narrow or too wide, droop, appear rounded, or overdeveloped muscularly. Figure 32 shows some of these variations.

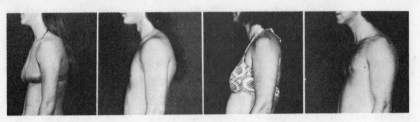

FIGURE 32. VARIATION IN FORM OF SHOULDERS.

Subject one shows the shoulders being pulled forward. This is always associated with some collapse of the upper chest, and may be seen as part of the constraint of chest emotions. In many females, with the rapid growth of the breasts during puberty, the various social taboos about them as sexual organs become strong factors in emotional development, and may lead to a drawing in of breasts, chest, and shoulders in an attitude of shame, embarrassment, and self-protection.

In some people, the shoulders are drawn upward as well as forward. These individuals may be visualized as withdrawing into themselves, somewhat like turtles. They are literally drawing their heads down into their shoulders. Psychologically, the same withdrawal is present, with fear, lack of assertion and a vague sense of impending punishment.

In subject two of Figure 32, the position of the shoulders is approximately normal. While in this person the line of the shoulder does not quite fall along the idealized lateral line of Figure 5 (page 28), he does not present any marked aberrations. Subject three, in the same series, is transitional. In subject four we see the shoulders clearly pulled back. This position is most frequently associated with a holding back of anger and a fear of striking out. When these emotions are finally released, they are found to be very forceful, even violent. The anger and rage in these people is also seen in the position of the jaw, which is thrust aggressively forward. One of the clues in evaluating the position of the shoulders is their relation to the head. In general, if the head is thrust forward markedly, the shoulders will be pulled back.

In another variant, the shoulders may be too wide for the general proportions of the body. This type of disproportion has already been discussed in the section on top-bottom displacements (page 43). The shoulders are overdeveloped, indicating that the individual is predisposed to perform the "doing" functions of the upper extremities. He is prepared to

FIGURE 33. WIDE SHOULDERS.

take on the challenges of producing and being a "man" in the world. When we meet with overly broad and developed shoulders in a woman, we often find that, on a deep level, she experiences herself as having to accomplish and take on functions that are most often, in our culture, relegated to the males.

On the other hand, the shoulders may be too narrow or slight. Here, the opposite situation may occur, with the individual lacking the development to take on the burdens of life. He may feel weak and unable to sustain his actions. His "doing" functions are impaired. This shoulder structure is frequently seen in people whose major complaint is not having

FIGURE 34. NARROW SHOULDERS.

enough energy to get things done. If the shoulder also droops, a more severe impairment of the ability to function is implied. Such people have great difficulty in taking charge of their own destinies.

The overly rounded shoulder (see Figure 9, page 39) is often associated with increased muscular development in this area. The underlying feeling is one of being overburdened or overloaded. Though the "weight" carried causes the body to bend forward (causing rounding), their bodies respond to this pressure by increasing the size of the shoulder and upper back musculature. This pattern is often seen in people who have great difficulty expressing themselves. It is as if their impulses are crushed by the same weight that can so easily be imagined as resting on their broad backs.

THE FACE

Of all the parts of the body, none is so directly expressive of a person as that complex unity of structure we call the face.* Just as the rest of the body expresses patterns of fixed emotional holdings, or energy blocks, so does the face.

Through a fluid, open personality may show, over a period of time, a vast number of different facial expressions, the expressions of an individual with emotional blocks will usually vary around some particular set pattern. If the block originated early in life, the structures of the face may indicate the same underlying emotion as the expressive pattern. The thin, tight mouth expressing bitterness or the narrow, intense eyes glaring suspicion are examples. Several easily recognized expressions are:

> wonder surprise anxiety fear panic terror annoyance blaming hostility anger hate rage longing sadness despair fatigue pain suffering cynicism contempt disgust innocence confusion excitement happiness joy contentment peace bliss

---

* This section on the face includes an integration of some of the teachings of A. Lowen, G. Gurdjieff, F. Peris, R. Stone and O. Ichazo. The basic insights of W. Reich and I. Rolf serve as a foundation throughout.

There are so many, no list could do them justice.

When the emotional life of an individual is explored, we usually find a central theme about which that life moves. One's world may be approached with a constant sense of terror, persistent suffering, fear of being reproached. An unchanging face is a mask we present to the world. Perhaps by appearing "stupid" or "innocent," we hope nothing will be asked of us. Or on seeing our "longing" or "panic," we expect to be comforted. Or in showing our "cynicism," we want others to experience our doubt and apathy regarding life and its energies. In some individuals, the facial mask becomes so organized that it takes on the nature of caricature such as: an "imp," "witch," "monster," "demon," "clown," or "devil."* The reader, with practice, should be able to recognize these caricatures in those about him, as well as him- or herself.

When attempting to analyze a facial expression, some questions we pose for ourselves are: Is the face, as a whole, alive and full of expression, or is it expressionless and blank? Is the face symmetrical or asymmetric? That is, does it have a left-right split? Is the skin healthy and glowing? Is the general shape of the face long and thin, oval or round? What are the various skin colors present? What do they tell us about the individual's biological and emotional status?

After the above evaluation, more specific parts of the face are brought to attention; we may divide these into four areas.

The face is naturally divided in this fashion. The nose and the nasol-labial folds serve as the dividing markers. Area 1 includes lips, jaw, and mouth. Area 2 includes the right side of the face from forehead to nasol-labial fold. Structures found here are the right eye and eyebrow, the right forehead, and the area beneath the eyes. Area 3 includes the same structures as Area 2, only on the left side. Area 4 refers to the immediate area between the eyebrows.

In Figure 36, Body Within The Face, we can begin to note some functional relationships of the face to the body as a whole.

---

* See *The Betrayal of the Body* by Alexander Lowen.

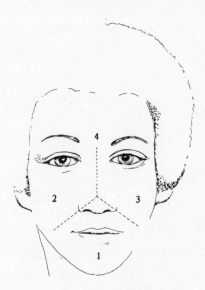

FIGURE 35. AREAS OF THE FACE.

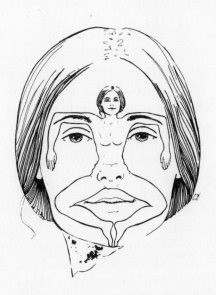

FIGURE 36. BODY WITHIN THE FACE.

TABLE 2

Area 1

jaw	pelvis
mouth and lips	genitals and anus
nasol-labial folds	under genitals and anus, thighs
chin	ankles and feet

Area 2

eyebrows and eyes	right shoulder, right arm, forearm and right hand

Area 3

eyebrows and eyes	left shoulder, left arm, forearm, and left hand

Area 4

forehead immediately between and slightly above eyebrows	head (brain—pineal gland)

We will now consider the parts in each area separately.

Area 1

The jaw may be overdeveloped and aggressively projected forward. Expressions related to the jaw include those of assertion, of getting what one wants, as well as eating, which aggressively involves actively chewing and biting. When the tightly held jaw is released by a deep tissue massage, anger and rage are often felt and expressed. It is interesting to note the relation of the jaw to the pelvis. The release of the pelvis is accompanied by similar emotions. In its sexual function, the mobilized pevlis is a vehicle for deep release and self-assertion. Blocking of the pelvis is associated with holding in the jaw. Hence, the jaw may be seen as the pelvis of the face.

Figure 37 illustrates the overdeveloped jaw. These individuals block and bind their emotional charge. The jaws may clench constantly, indicating ever-increasing charge which,

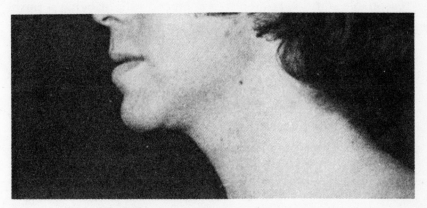

FIGURE 37. THE OVERDEVELOPED JAW.

when lacking discharge, causes overdevelopment, Hence, the large jaw and jaw muscles. Feeling the pressure of blocked energy, these people project, with their jutting jaw, an aura of extreme aggression, and great determination. They are, in some sense, in a double bind, for their held energy cannot find an outlet. (Recall the aforementioned pelvic sexual block.) And so, they are internally driven. This may manifest itself on a social and interpersonal level as "pushiness," or a tendency to accept any challenge. However, these outward-directed actions are inadequate to release their bound energies. What is more, by acting aggressively they provoke an inner aggression which leads to more energy binding: a vicious circle exists. In therapy, the individual hopefully is able to gradually release his held-in anger safely and to an ever-increasing degree. With time, the jaw muscles relax, releasing the jutting jaw.

Alternately, the jaw may be underdeveloped, which is commonly referred to as a "weak" jaw and associated with a lack of assertion. We find here that the individual is most often undercharged. He is unable to take what he wants, unable to "get his teeth into it" or to "get a hold on it." In the authors' experience, a normally well developed jaw appears to be related to a well developed will. It may be noted in Figure 36 (page 91) that the chin and feet are related. In the same way, taking a reasonable stand is related to a balanced will, or being a push-over to a weak will.

The shape and activity of the mouth and lips tell us a great deal about an individual. It is widely accepted that the earliest stage in human development is centered around the activities of the mouth, especially sucking. When, for whatever reason, the infant's gratification at this stage is repeatedly interrupted, he experiences frustration. No matter how he may try to obtain satisfaction, to suck or cry or move, that is, *To Act,* he fails. When this pattern of frustration is reinfroced by constant parental injunction ("You can't do anything right") from infancy through childhood, the individual comes to experience a deep-seated insecurity regarding his ability *to perform.* That is, in attempting to progress from passive dependence on the mother to independent action, he is frustrated. The child may be asked to do things beyond his capacity, or he may be held back, for example, if he attempts to feed himself but is too messy. He may be overprotected and not allowed the experience of succeeding and overcoming difficulties. As an adult, we will find these individuals traveling through a life pattern in which they are repeatedly attempting to discover the most appropriate manner in which to act. They are constantly vacillating.

Sexually, a similar pattern occurs. Many partners are experimented with in an attempt to find out "how to perform" in sex. Or then again, if there is sufficient insecurity, the individual will hold back from acting at all. He or she may be the wallflower, watching everything happen from the outside. In effect, these individuals are simply "out of tune." They don't know what appropriate action to take. They may attempt to force their "tune" on others, or they will often act in a noisy, loud fashion, their mouths going on and on. A prototype of this group is Mussolini. The Duce's curled lower lip was part of his trademark. Many of these individuals retain strong oral traits.

Oral Traits Related to Food as an expression of Emotional Deprivation in Infancy: Some overeat and become obese. Others under-eat or, being gluttonous, will eat a myriad of mixed foods. Many activities involving the mouth are noted; chewing gum, lipstick, smoking, pills of all kinds. Anything and everything is indiscriminately shoved into the mouth. As we mentioned, these individuals often talk excessively, loudly, and innappropriately, since they are "out of tune" with their environment.

Oral Traits Related to Dependency: As mentioned, this aspect is, for the authors, of singular importance in this group. The infant, initially passive and dependent on the mother, begins early on to strive for autonomy. The healthy person can enjoy having things done for him, while remaining at the same time confident of his own ability to take independent action. This ability to act, if markedly impaired, results in the adult character we are describing.

Oral Traits Related to the Experience of Insecurity and Discomfort: When security and comfort have been repeatedly threatened during the initial period of growth, the individual begins to develop ideas about how to act on the environment so as to render it secure, consistent, and comfortable. His ideas are either totally idealistic and ungrounded (out of tune), or dogmatic and unmoving.

Oral Traits Related to a Lack of Tenderness, Control, and Affection: During the early stage of development, physical contact as an expression of affection and tenderness may be impaired. The individual experiences a sense of lack. As an adult, he attempts to achieve contact and warmth through sexual interaction. However, not knowing how to act appropriately, besides lacking an inner sense of security, he will engage in frantic, indiscriminate sexual behavior, or withdraw into sexual timidity.

The tendencies toward withdrawal may become predominant, and the individual will present character traits to be found in the next stage of emotional development, namely the anal stage. This stage is involved principally with the attention to bowel and sphincter function. How toilet training has been handled helps to determine the individual's attitude toward punctuality, authority, neatness, doing one's "duty" every day, and, finally, holding on to things. That is, the feces may be seen as a material possession to be held onto even in the face of extenal parental pressure to use the "potty." The most significant trait here, however, seems to be that of "holding onto things." Such individuals may be found to be avaricious. They are collectors of stamps, goods and, most often, money. They are stingy with material goods as well as with giving of themselves emotionally. The are emotionally withdrawn and insecure. They hold back, not daring to commit themselves, since they really do not know what that means. Instead, they hold on, keeping everything to themselves.

When the tendency to dogmatize and/or idealize takes on prominence, a character type develops that generally appears to include the following:

1. Many grandiose (overblown) ideas.
2. A dogmatic defense of the faith, whatever the faith may be: America, religion, capitalism, communism, etc.
3. Overweight (or a history of being overweight).

All of the above, as we have already indicated, can be seen as an expression of not knowing how to act in order to accomplish the gratification one desires. Orality is most clearly expressed in "mouthing" many ideas about "how one should act" to set things right (so as to attain gratification).

Hence, upon viewing a face, Area 1 should be examined in terms of the predominance of the activity of the mouth as compared to Areas 2 and 3 (eyes). Is this person constantly "shooting off his mouth?" Or are there areas of his mouth which lack life and activity? The mouth may be retracted to one or another side, or a person may speak out of the side of his mouth, or again, like Mussolini, project the midlower lip forward.

It is beyond the scope of this present work to go into further details. To summarize, the shape and activity of the mouth makes a direct statement as to the degree of orality present in the individual. Our central concept is that during the oral stage, normal development allows for feelings of security which enable the person to grow, self-assert, and discover. When this process is disrupted, the result is a feeling of insecurity, as well as a basic inability toward action.

In this illustration, the mouth is drawn to one side in a typical distortion. It should be easy for the reader to imagine

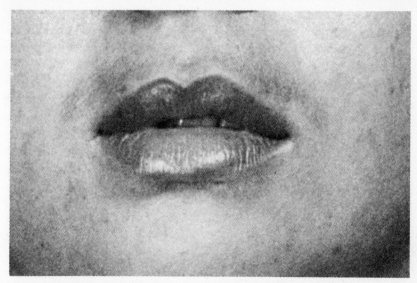

FIGURE 38. MOUTH WITH RETRACTED RIGHT CORNER.

the mouth drawn to the other side, or with the central area of the mouth pushed forward.

Delineating Area 1, and separating it from Areas 2 and 3, are the naso-labial folds. Please refer to Figure 36 (page 91). It will be noted that the thighs occupy the grooves of the naso-labial folds. In our experience, those individuals who present deeply drawn folds in this area have lived lives which continuously drain on their energies. It is a sign of depressed vitality and strain. In a sense, their ability to take action has been exhausted.

The thighs, esoterically, have been related to capacity: That is, what an individual can do in terms of available energy. If the reader will take a moment to stand up, shut the eyes and place awareness to the thighs, he or she should be able to feel the capacity of energy within.

Areas 2 and 3

The most prominent structure, of course, is the eye. The left and right eye have different energetic, emotional, and personal-historical significance.* In some esoteric schools and Western growth communities, the left eye is referred to as the eye of "essence" or "being," while the right eye is the eye of "personality" or "ego." In terms of our work, the left eye reflects more the integrity of the "core" or "intrinsic being" layer (see p. 100), and the right eye reflects the "extrinsic" or "doing" layer. The left eye is the "receptive" eye. Through it we take in the world. In this sense it is Yin or female. It reflects the nature and character of our earliest life relationship, namely that with our mother. If this relation was a healthy one in which we felt nourished and secure, free of all fears of abandonment, then our sense of self or beingness is solid. In effect, we become "somebody," a person with a secure sense of our own existence as a person, by virtue of having been treated as such by our mothers. We know who we are and are secure in our self-worth and self-esteem. If the relationship was not nourishing, then we lack a sense of self. Our perception of ourselves and the world about us is radically different from those of a person with a strong sense of his or her own being. Anxiety and fear of abandonment**stains all our experience. We see the world through a screen constructed of the emotional charge we had with our mother.

When viewing the face, the balance and energy between the two eyes is evaluated. In Figure 39, we can clearly see that the left eye lacks the life and vitality present in the right eye. We can conclude that this person lacks a deep sense of self-worth, is anxious, and has difficulty in receiving. Fundamentally, he doesn't know what it means to be simply present. He elaborates and, as a reaction formation, tells others how to be. Often, he assumes religious, philosophical, or spiritual leadership roles. He searches with-out himself. Feeling worthless within, he frequently becomes apathetic. On the other hand, he also

---

* In the field of homeopathic medicine, inherited disease traits occur in the left eye when inherited from the mother, and in the right eye when inherited from the father.

** Karen Horney, a neo-Freudian analyst, stressed this concept of an early "basic anxiety" as underlying all neurotic behavior patterns.

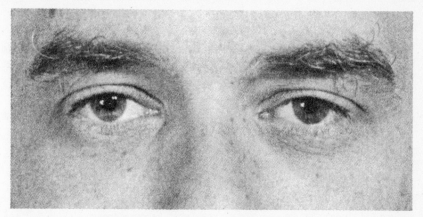

FIGURE 39. LEFT EYE LACKING ENERGY.

constantly attempts to find something or someone that will give him a sense of worth and beingness.

The right eye, as the eye of personality (ego) or extrinsic doing layer, reflects our relationship with our father. It is our active, Yang, masculine, outgoing eye. Through it, we project out energies. If our relationship with our father was secure (in a sense, if he took us by the hand and showed us the way in the world), then our ability to deal with our relations to others is healthy. The father, after all, adds a third important dimension to the child-mother bond. In dealing with this relationship, the father lays down the pattern with which the child will relate to "others" in general. A disruption of this pattern leads to numerous relational problems. We see the world through the screen of a distorted child-father relationship. Not knowing how to relate, we may develop paranoid systems, in which our will is being tested. There is an ongoing, deep-seated distrust of others. Often, to overcome this, these individuals will attempt to take over and run the show. Not trusting, they feel the need to control. Not knowing what God or society's trip is, they decide that it is best to become "a little god." This is reflected back to the world as vanity, accompanied by some degree of aggression. Homosexuality may be present in this group of individuals, since they have a relational problem with the father. Not trusting others, they will often present a facial mask or screen to the world. Hence, if a lack of life is noted in the right eye, then we know that the energy of the child/father relationship was impaired with all the adult attitudes that we have described above coming into play.

What follows is a table summarizing some of the basic ideas of the preceding pages:

TABLE 2. AREAS 2 AND 3 SUMMARIZED

Right Eye	*Left Eye*
personality (ego)	essence (core self)
relationship with father	relationship with mother
social relationship impaired	sense of beingness impaired
distrust and paranoia	anxiety
extrinsic layer	intrinsic layer
doing	being
active (Yang)	receptive (Yin)
male	female

Above the eyes, the eyebrows and forehead also express mood and character. The eyebrows may raise in surprise, puzzlement, fear, pain, or terror, all creasing the forehead. In keeping with the body-face parallel depicted in Figure 36 (page 91), the shoulders, like the eyebrows, are raised in fear. The creases of the forehead, especially when deep and permanent, point to long-felt cares and troubles, fear for one's loved ones and way of life. The smooth brow is most often associated with feelings of inner peace and an untroubled existence.

Area 4

A fourth area, the area between the eyebrows, is also of great significance. In Figure 36 (page 91), it can be seen that the head or seat of mind consciousness is located between the eyebrows. This area has long been considered the site of the third eye, or the eye of wisdom.* When the area is knotted together with ridges of tension, this is associated with a feeling of intensity and/or bound anger. The reader is invited to knit his brow and experience any sensation that arises. A person who has a persistent contraction of the brow may be seen as narrowing his vision or focusing attention upon a limited field. When angry, we are unable to see very far beyond our own immediate feelings.

THE HEAD AND NECK

The following series of silhouettes illustrates the varying positions of the neck and head in relation to the whole body. The figure in the center approximates the idealized normal. The head and neck occupy a position over the center of a line drawn through the shoulder. In the first subject the head is thrust forward of the body. This is often found to be associated with a driving, aggressive attitude. It is, in a way, the foremark of contemporary man. He uses his head or brain as his prime means for planning what he wants in the world. In the third subject, the head is seen to be tilted back with a shortening at the base of the skull. Great tension is found at this location. Obviously, this person is holding his energy back with his body and head. Release of the tight musculature of the back of the neck by massage is invariably accompanied by strong emotional discharge. Accompanying this tightness of the neck is a tightly drawn scalp. The tightness extends from the base of the skull over the top of the head to the face. When the tension at the base of the neck—that is, the junction between the base of the skull and the neck—is released, it is usually accompanied by a release of tension in the jaw. Any

* In esoteric literature, the third eye has been associated with the pineal gland. This gland is found in the center of the head between the two cerebral hemispheres. Its functional activity is regulated by light coming through the eyes. One of the substances isolated from the gland is known as "melatonin." It appears to have a regulatory function in terms of sexual development and activity. The secretions of the pineal gland, since they are light related, may serve as an internal clock to the body in its diurnal cycle.

tension found in the eyes (this is common with tension at the base of the skull), is also released by manipulating the neck muscles. Eyes which were dull and lifeless, or sharp and hard (more common with this posture), are seen to soften and light up.

In addition to the forward and back tilt of the head, there may be an associated tilt of the head to the right or left. When this tilt to one side is present, the individual, in a sense, cannot approach anyone or anything directly. He is not being "straight." This particular posture of the head is often found in individuals who may have an associated disorganization in the relation of the various body segments. This, as we have already mentioned, is present in what we've called the schizoid or schizophrenic character type. To come toward a situation in a straightforward fashion may be more than the individual with a disorganized body can handle.

The neck, of course, may be very long or very short. With a very long neck, it often appears as if the subject is split between his head and his body. In fact, many such subjects do present an emotional-psychological state in which the body and mind are separated or disassociated. In other subjects, the head jams the body with a very short neck, almost as if the subjects were contracting and withdrawing into themselves.

We wish to remind the reader again that, while this type of dissection of the body into segments is useful, it is also limited. Only contact with the whole man or woman allows us the possibility of truly understanding what each single aspect of the personality means. In a sense, while it is true that each part reflects the whole, it is the dynamics of the whole that determine the particular meaning of the individual segments. A jutting jaw or pulled back shoulders may indicate underlying anger or rage, but it is only by contacting the heart of the individual that we can hope to know him and what his rage is all about. We caution the reader against establishing any set of easy formulas.

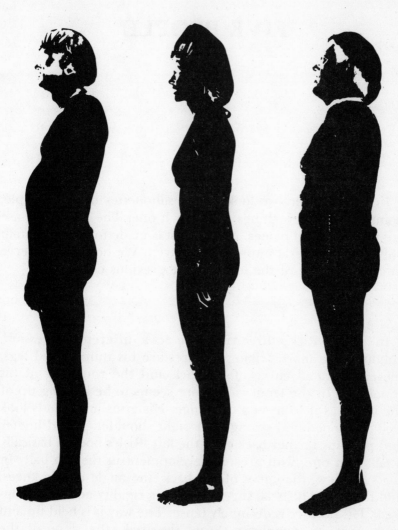

FIGURE 40. HEAD AND NECK.

5.

FIVE PEOPLE

In this chapter, we look at the silhouettes of five people, pointing out a few things about each one. The names we use are not their real names. No attempt is made to go into detail; only a very general impression is given. We hope it will serve to ground some of the ideas in the previous chapters.

RICK

In each of Rick's three views, we get a different impression, although the most striking features are his thin, bowed legs, the exaggerated curve of his back and the rounding of his shoulders. In the front view there seems to be a rising up of the body, as if to meet a challenge. His arms and hands look well developed and strong. The right shoulder is held higher and perhaps farther back than the left. Rick's body is basically a tall, thin one, with greater development in the top half. In the side view, the curve of the back, the angle of the knees, and the raised belly all suggest a strong rigidity and a holding-back. His posture is definitely tense. The way it is held up and back suggests immobility. From the back, the slope of the shoulders gives an impression of resignation or of questioning. The contrast with the front is most striking. Viewed as a totality, there is an upward shift, a slight left-right split, and an overall rigidity and immobility.

We might speculate that the weakness in Rick's legs is compensated for by the development of his shoulders and arms, and the rigidity which attempts to keep the whole structure up. The mixture of elements makes Rick's feelings difficult to read. But that in itself is characteristic.

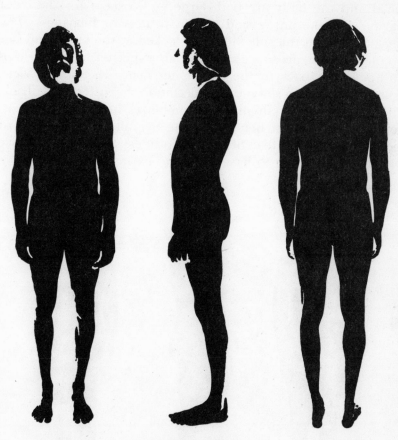

FIGURE 41. RICK.

ELLEN

Ellen's body shows compression. It is crushed down from the top, and drawn in at the waist. Her head is tucked down into her shoulders, which are held up tightly and drawn in toward the ears. Her buttocks seem drawn in and up. The head and the body are both large and, except for the chest cavity, give the impression of massiveness or heaviness. The legs, as seen in the side view, are locked at the knee, and her whole body seems braced against an expected blow, perhaps to her upper back or the top of her head. In the side view, Ellen's hands also show this expectation. Seen from the back, the tension in her shoulders immobilizes her arms while attempting to protect her head. The overall picture is one of holding in, with a fear of, yet a readiness for, an attack. We would expect Ellen to have difficulty expressing herself.

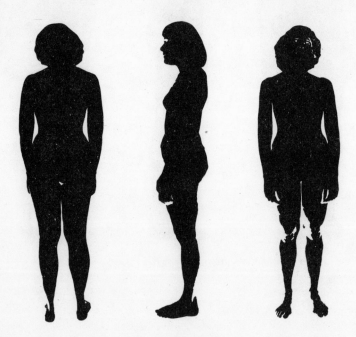

FIGURE 42. ELLEN.

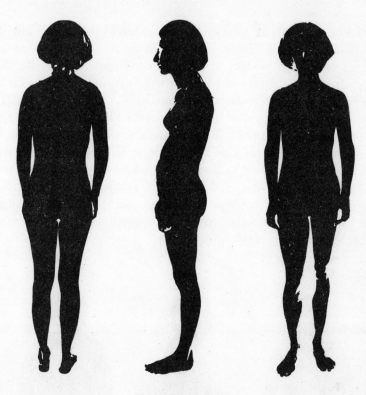

FIGURE 43. FRAN.

FRAN

In both Fran's front and back views, there is a sense of
expectation, illustrated by a slight forward movement of the
left leg and a drawing back of the right arm. In the side view,
however, the feeling is one of collapse. The knee is locked, the
pelvis tilted, the belly out, the chest collapsed, the shoulders
and head forward. Except for the knees, which are locked,
Fran's body seems to be folding up, indicating a low level of
energy. Behind such a posture would be a general weakness
and a need for support.

LOUIS

In Louis, there is a large displacement upward. While the top half looks full and strong, the legs are knock-kneed and seem weak, with widely splayed feet. Especially in the back view, the legs and feet convey an impression of dismay, a lost or confused feeling. In the front view, Louis's top half is raised up and expanded as if he were trying to overcome the lost little boy feelings of the legs by a strong effort of will. In the side view, the rounded shoulders, the bend of the back and the tucked in and pulled up buttocks show a compression similar to Ellen's. We would expect Louis to have difficulty expressing himself, except for occasional outbursts of anger. Due to the displacement upward, an emphasis on fantasy and mental activity are also quite likely.

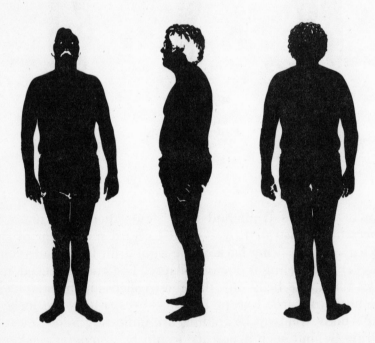

FIGURE 44. LOUIS.

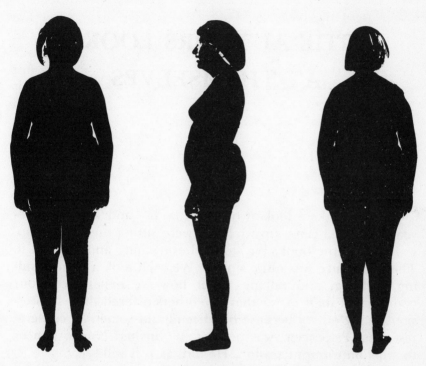

FIGURE 45. MARY.

MARY

Although Mary is a woman in her middle twenties, her over-all body is like that of a child. The narrow shoulders, small arms, hands, and feet and the nearly formless lower legs seem to belong to a very young, weak little girl, almost an infant. All her energy is held in the large area around the pelvis. From the side, we can see that the back breaks high, very sharply, and that the pelvis is strongly retracted and held. The impression is of someone stuck, unable to act or move. From the back, helpless, forelorn, and slightly anxious feelings predominate. There is strength and energy here, but it is lying dormant.

6.

THE AUTHORS LOOK
AT THEMSELVES

Here, we take a look at our own bodies and what they reveal. The idea came up while we were sitting out in the sun, working on the book. We stopped everything and just did it. The procedure was quite simple. We each took a turn standing, walking, and talking about how we experienced our bodies right then. After that, the others offered their impressions. I say others because our friend and sometime cotherapist, Sam Pasiencier, was present. We invited him to join us for this impromptu session. He did, as you will see.

We had a tape recorder going to capture the dialog. With a few improvements in the grammer and syntax, it is reproduced here exactly as it was recorded. Summaries, some explanations, and the necessary stage directions have been added for clarity.

Hector went first.

Hector: I'm standing in the sunlight and as I feel my body, I feel first my legs, my feet pressing on the ground. And I feel an increase in the weight in the right foot and the right leg. And I feel a rotation, an inner rotation of my body around the left. I feel my right shoulder in front of my left shoulder. I feel tension in the back of my neck. Not as great as I have felt sometimes, but there is some tension there. I feel a little more tension up in the right shoulder. I feel tension in the lower back. I feel the lower back arching forward. I feel some strain

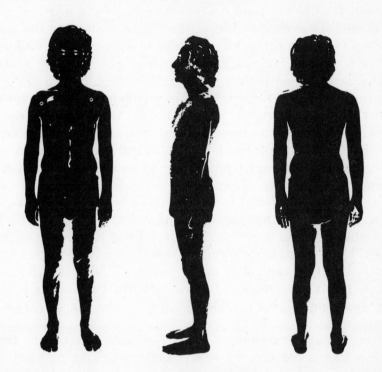

FIGURE 46. HECTOR.

behind the knees. I feel pretty good in the buttocks. I don't
feel too tight. Not as tight as I've felt in the past. I feel my
shoulders are not free. They don't feel free. They feel
blocked. They're holding in some way. This is a subjective
experience, that they're being held. The hips. I remember a
time when I felt really pinned in the hips. They're not as
pinned now. I feel my breath go into my belly, and some of
the tension I've been carrying around in my belly. In my face,
the thing I'm in contact with is my squint, in my left eye, a lot.
I feel the . . . my sensation of my face, it is drawn. The sensation in my face is as if it were drawn downward. The energy
is coming downward in the face. And my face feels like a long
angular face with a downward direction in the front of it. It

feels discordant with what's going on in my central nervous system. It's like my central nervous system feels alive and my face feels relatively down. But I feel light coming out this way (*the right eye*). I'm much more mobile on my right side, and my left side is hanging much more passively. My weight is not really balanced on both legs. There's some shifting that's going on. And I feel fairly relaxed in my body, except for these areas of tension here, the lower belly, the tension in the lower back, some tension up here (*the shoulders*), and I think the face thing is really important. I feel that. And some of the inner streamings and things I feel. I don't feel as blown-up in my chest as I used to feel. There was a time when I felt . . . (*Hector holds his chest expanded.*) I used to be this way, as you say, a rigid type. I was very much this kind—retracted pelvis, held back. But now I feel a lot of that changing. I like the feeling of my chest. I feel much more receptive than I ever felt before. I feel much softer in here and much easier. Though I'm feeling these strains I indicated. The low back, the back of the knees are straining. There's some shifting of the weight. That's about it.

Ron: I'd like to comment. When I soften my vision, the first things that come are the really solid feet. They're planted. They're solid. And there's a slight tendency to carry the weight on the outside of the calf. The whole body seems very soft, soft lines. The head and the neck seem a little big for the body.

H: Right! I feel that.

R: It's soft. And there's a little dip. There's a little longing in there (*the chest*).

H: There's a little caving in of the breastbone. Feel's like it has to do with this kind of fear and this kind of collapsing in here (*folding forward of the body*) of . . .

R: . . . of suffering.

H: Suffering. Right! Suffering.

R: The arms are a little . . . you're right . . . the shoulders seem to be not quite connected, not quite . . . they're a little frozen, you know.

H: It feels that way.

R: The softness ends right along the top of the chest. And the shoulders are back a ways. They're held back a ways. The head is out a little.

H: There's still some of that tendency.

R: Yeah. The head is out a little.

Sam: I see a lot of crossing. I see a left angle up to that shoulder.

H: I feel this twist. Is that what you're seeing?

S: Sure.

H: So I'm twisted. On some levels, I'm twisted, my personality, my being is somehow twisted. Let's not say my personality. Let's say my being. It has a twist in it. Which goes with the eyes. Some twist. You know I like that a lot better, experiencing the twist. I feel it. And I think you can see with the foot, Ron, the left foot is out.

R: Oh, yeah. Yeah.

H: It's turned out.

S: You can see the pattern of tension in the body. The left leg is tenser than the right leg.

H: I feel that.

S: You tensed that up because I mentioned it.

H: No, no. I feel that difference. I wonder if this has to do with this (*a punching motion with the head turned away*). As if it were a swing this way and a getting away. Like I don't even want to watch the swing.

R: It's like that (*imitating the swing*).

H: So I don't want to watch it.

S: The head is turned away. You don't look.

H: So, I don't look. Right.

R: But you're putting a lot of energy into it. The energy is coming all the way from the ground.

H: I think it has something to do with that. You know, there's a striking part of me that I don't even want to see. That's my subjective interpretation.

R: Yeah. That's right. I feel that. I feel that . . . There's this in your face (*a sunken-cheek look*). It's sallow.

H: This is the . . .

R: The cheeks are a little in . . .

H: Undernourished, undernourished.

R: There's a suffering in your face, a real suffering.

H: A lot of suffering. Suffering in the face, in the jaw. Lack of nutrition that I feel that I want. Some more nourishment.

S: I experience a lot of openness, very little shame.

R: And there's a feeling of being a good boy, too.

H: Right! Right! I have that, definitely. I want you to like me . . . How about this lower back here? How much tension do you see in my buttocks? Cause I've worked a little bit on that.

R: Well, there's a little bit, a slight tendency to collapse in here.

H: Where is that?

R: Here. And there. (*Ron is touching Hector's back near the waistline.*)

H: I think I have some oral traits.

R: Do you feel tension here (*along the spine*)?

H: Not the way I used to. I've let go a lot of the tension here. Well, I feel an orality about me, too.

R: From the back, you know, if I wanted to make a fantasy, from the back there is this good boy, but then he really wants to punch, too, to give a good punch.

H: Right! I think the good boy has to do with my need to be fed.

S: There's a dimunition in the sides of the legs. I think your legs are small in relation to your body.

H: Oh, right! And then I want to make one more comment about what I experience, which is a lot better after rolfing and all the work I've been through.

R: The space between?

H: No! Between the knees . . . the tightest part of me has always been between the knees and the ankle. You see this area here (*between the right knee and ankle*)?

R: George (*George Simon, a friend and teacher*) picked that up instantly. Remember that?

H: Yeah, he picked it up on me instantly.

R: "You're being inflexible," George said.

H: And that's it! I feel this tightness a great deal. It's as if my legs are kind of like . . . stilts.

R: Uh huh. The arm even got straight and stiff.

H: There's a rigidity here (*the arm*). I'm, of course, exaggerating that. But that's the feeling. There's a certain inflexibility. . . .

R: Would you put your feet together please, Hector?

H: There's a certain inflexibility between the knees and the ankles.

R: Yeah. It's between the knees and the ankles. The thighs come together beautifully. You also have a tendency to this curve, where the lower back comes in a little.

H: Well, that's part of that rigid thing. Left over. It hasn't been worked out.

R: You also have a little . . . it's almost like . . . it's almost getting a crushed look in the chest, if you kept going.

S: Rounded shoulders.

R: If it starts to go. See, keep your head there! The shoulders are starting to fold over.

H: That's the pain and suffering in here (*the chest*).

S: The spine doesn't show it too much.

H: But it's here (*the upper chest*).

R: It's starting.

H: It's this foldover.

S: It starts here. The spine . . .

H: It's holding the flexors. I also experience I don't want to give out my feelings, too. Even though, with all the openness that I've had, there's also a deep layer of not giving them out.

R: Ah! That's it!

H: In other words, there's a surface layer where I can be open, right, but you come in and touch me deeply and I have to work through it. It's still there.

R: You mentioned the jaw, which is broad . . .

H: Yeah, the jaw is broad, I feel the broad jaw. I feel that. Then my eyes. I want to say something about my eyes. My eyes always used to put people off a great deal, because there was a sort of that rigid thing, grrrr, ready to strike them. And people would say to me, you know, "I'm afraid of you." I'd say, "What? I don't understand that. Why are you afraid of me? I didn't do anything to you." Right? And I feel that's let go.

R: The eyes are bright and shiny.

H: Quite a bit.

R: The right eye is shining.

S: You want to try something? This is an idea.

H: Oh! One more thing. My dominance is definitely right-sided. Right arm. Right hand. I'm split right-left. In other words, my relational thing is this thing *(twists toward the right)*. My nourishing thing, the mother thing, is still screwed up. And in fact, even in Oscar's system *(Oscar Ichazo, founder of Árica)*, I came into the being group where the relation with the mother was distorted. And I feel that in my body. And, in fact, my strength is that way. We said the movement is to the right. It is. It's not this way *(to the left)*.

S: Mmm. I was going to ask you to just walk. Walk around! *(Hector starts walking in a large circle.)* And maybe make slightly large movements. More than just walking. Any kind of larger movements. Anything you like. *(Hector swings his arms up and over and forward and takes small jumping steps.)* And then come to a stop! *(Hector stops.)* And now experience your body!

H: The sense of motion is very full. Like I'm into action. I'm an action person. I feel that. Action for me is very, very much a part of me.

R: Hector, when I saw you walk, the things that I saw immediately were the shoulder thing again and a little bit of this here *(the lower back)*.

H: Those things are still here.

R: Walk again! It's held right back in here . . . *(the lower back)*.

H: That's where I thought there would be holding. I haven't cleared my body yet.

Before going on, we'd like to summarize the work with Hector. He'd been a somewhat rigid person before starting to work on himself, say six years ago. And though he'd come quite a long way, some of the rigidity was still there. It was in the calves and shoulders and lower back. His head was out a bit and the shoulders were held back and somewhat immobilized. These tension areas were reflected in the feelings he withheld: anger, for one, which was also seen in the twisting motion as if to strike out; and fear, seen in the looking away, and almost certainly connected with the "good boy" posture.

Another dimension lay in the suffering and need for nourishment. The smallness of the body and the beginning of the folding forward of shoulders and upper body fit these feelings. But there are some good points to mention. The feet are well planted and strong. The whole structure is generally flexible and soft. The eyes shine. These good points manifest in Hector's openness and energy, his ability to function well in reality, and his wide range of interests and activities.

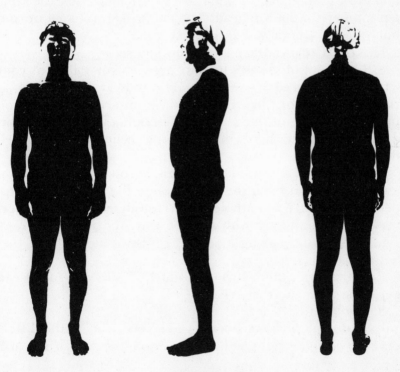

FIGURE 47. RON.

R: I feel tension starting from my left ear running down right past the scapula and the shoulder and into the left arm. This arm (*the left*) feels frozen. Dead. O.K. There's tension right in the middle of the brow. And a general feeling of rigidity. A slight feeling of tension in the abdominal wall. The deadness in the arm is a very prominent feeling. There's a slight tremor through the legs, deep in the legs. I also feel right now . . . a kind of shifting relationship to the ground. My feet do not feel well planted right now. I'm breathing all the way from here (*the lower belly*), but it's shallow. The breathing feels shallow. My legs are vibrating inside. My weight seems to be . . . on my feet. Tensing around the buttocks a little. Now I want to change my posture. I sense that whole thing about defiance, will, struggles of will. I feel that energy. That makes my body feel alive when I get into that. I really could fight now. The energy level is not bubbling and it's not collapsed. I don't feel tired . . . and I don't feel excitement. I feel the same tension here in the forehead. The energy seems to get up and hold there. Generally, I feel pretty good.

H: O.K. Go ahead Sam, what do you see?

S: I see a lot of sadness in your eyes. As you talked of your left arm, I noticed it's pudgier. The muscle doesn't look as functional. It looks like this could be an area in which you turn to the kinds of things you're describing in the book. This would be an area where you're holding feelings. And the feeling is impacted.

R: Yeah. And I think it runs right through the whole system. Right down from the ear, through the arm and the upper back. It's like this whole part is paralyzed. And I do feel very silly doing things with my left arm. My right arm . . .

S: How about making a really aggressive motion with that left arm.

R: But I don't think I can get much . . . (*Ron tries it.*) I can get some power in that. Yeah.

S: Contact that!

R: It doesn't feel like I could . . . it feels like . . . the difference between the right and the left is that the right is me, and

I could hit with it, and I have no trouble with it. This *(hitting with the left)* is like using a weapon. This is like having an inanimate object in your hand to hit with, it's a tool, you know, it doesn't feel the same fluidity as the right one.

S: The lower leg looks very tense. The calves look very tense. The feet look small for the size of the body and very compacted also. Filled in.

R: They're slightly collapsed, I think.

H: Some of the things that I see. Ron, get your head up! Well, I see a duality in your expression and in your eyes. The duality is a striking out, rage almost, yet there's a soft, receptive area. So you've got both of them going. I feel that ambivalence. Now, if you look at your stance, even right now, your hands are up there. *(Ron has his hands on his hips.)* You're ready.

R: I'm ready to fight.

H: You're ready to fight. You're ready to swing. It's the beginning of that. So, that's in there. There's both the softness, that comes in the chest, and the collapse in the chest, which makes for some softness in you. The thing that strikes me most and beyond that, beyond the head and that aggressiveness, mixed with softness, is that in the legs, especially from the knees and up to and including the buttocks to the lower waistline back, there's a cutoff, there's a split in the back. There's a definite split right there between the thorax and the lower part of the trunk, and that split is reflected in front between the diaphragm and the belly. So this is all blocked energy, not flowing, from the thorax, from taking in, to the belly. O.K. So this is held energy. Now, the energy holds I experience in your case, from the ground up, in other words, your pushing the energy up from, if you look at the knees, the inside of the knees, from the knees to the perineum,* you can see that he's short. This is a relatively short area. And so the buttocks, as you look around him on the back, here it's even clearer. Sam, come and take a look here! You can see that it comes to this area under his perineum, where his buttocks are pinched and held.

S: You have a hard time letting go of this area.

H: This whole area is held. You should loosen up the perineum. That'll let out a lot of stuff. So I see there's blocking in at the hips. There's holding in here at the hips and in the inner groin.

R: That causes this thing to back up. *(The tension in the groin causes the belly to protrude).*

H: And that causes this to block up *(the buttocks area).* So there's three levels of blocks. Well, there's more, because the neck. There would be the neck. There would be the thoracic-abdominal block. There's the abdominal-pelvic block, and there's the knee to the groin block that's happening.

R: Sounds pretty schizy to me.

S: No, no. It's all the expression of that one communicative . . .

H: I mean that's what I'm seeing. I mean I'm seeing that those are the large segmental areas of holding.

R: I've felt those bands.

H: You've felt those bands.

R: I've felt a band across here *(below the belly),* across here *(the chest),* even the temperature is different.

H: I don't think you're schizy.

S: If there were radical differences in the character of the segments, then it would be schizophrenic. What he's saying is that's where the blocks are.

R: O.K.

S: He isn't saying you're disjointed at those points.

H: No, those are the areas of the blocks and, I think, the route to your opening; the route to your opening would be this opening, the pelvis. Well, first I would work *(rolf)* in this area *(the legs)* and this area *(pelvis).* And I think if that perineum were opened, naturally, gradually, 'cause he's got enough rage in him to really go through a trip with that. But I think that that opening would unfold the rest. That's my hunch.

R: When I feel it coming on, it comes on right here *(lower belly).*

H: I think there are a lot of nice blocks that we can show in the book. *(Laughter.)*

R: Why don't you watch me walk.

H: Does that feel right to you, the comments?

R: Yeah, it does feel right. *(Ron starts walking.)*

H: Well, what you can see is that when he walks, he wiggles. Which means that the front-back thrust is being impaired. And its' being impaired from the level of the break between the trunk and the pelvis. The trunk and the lumbar spine, rather. In other words, if you walk and you're wiggling from side to side . . .

R: I do?

H: Yeah. See his wiggle?

S: Yeah. That's what you call an "I can do anything" strut.

Hector: Well, I don't know what you call that, but that wiggle . . .

R: Oh yeah, I see now. I walk like this . . .

H: Right! So that wiggle, you see, is the reflex of that block between the thorax and the lumbar spine. Otherwise, you would be able to move forward and back. And the pulsatory movement would be happening that way. And it's not happening that way. It's wiggling around it. And there's a very definite block. You can see the block right there *(the lower back)*. It breaks high. That's what I see, Ron.

R: You think I can write a book though.

H: I think you can write a book.

S: You can write three books. *(Laughter.)*

H: That's nice. So what do you say? You say guys like Ron are holding themselves . . .

R: They're holding themselves upward.

H: That's what you have.

R: My head should be small for my thorax.

H: It is. Of course, we didn't comment about that. We should say that. Ron's head is relatively small compared to the rest of his body. His chest is expanded compared to his pelvis, which is narrowed down and tightened.

R: A classic displaced type.

H: Is that a classic displaced type? Well, that's it.

There are some splits in Ron, notably at the shoulder and in the pelvic area. The shoulder split results in the arms feeling cut off, and the pelvic split isolates the rest of the body from the ground. This isolation from the ground is also an aspect of the upward displacement. The upper arms and the thighs are short and pudgy, another indication of the dissociated feelings there. The general picture is blocked anger and a tendency to blown-up ego. There's a hurt feeling in the chest area. It should be noted that the heaviness is another indication of blocked feeling. So we have a picture here of a man who is holding back strong feelings, mostly rage and sadness. He contacts his world with his head. At least that's the tendency.

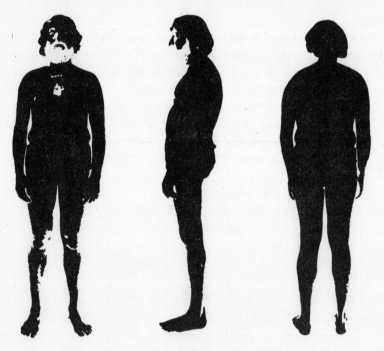

FIGURE 48. SAM.

S: Just take a little time to sense myself. I have a lot of . . . I have a very hard time sometimes just sensing my own presence. That's very frightening. In terms of the tensions that I feel immediately, I feel a throat tension.

H: Let your cigarette go, Sam!

S: Throat tension in the front. I had a belly tension, I jsut let it go. 'Cause I've gotten to the point that when I feel them, I can relax them. A lot of tension up the backs of my legs. In fact, my entire back . . . this is after a lot of work . . . but I feel tremendous tension across the whole of my back, from the heels on up. I feel my feet pretty much on the ground, but I feel that it's the backs of my legs that keep me from letting myself go to the feeling of my . . . gravity, my own weight on the ground. A lot of tension in the pelvis. Deep in this area, very deep in there *(the pelvis)*. Across the back of my shoulders I can feel shivering going down the back of my body. I'll tell you, the prevalent feeling across the back of my body is a total contraction. It is as if all the musculature were contracted. I feel very massive. Kind of bowing, here. This part *(the chest)*.

H: Bowing.

S: It makes me want to bow.

H: Bow forward. That's what it makes you want to do?

S: Yeah.

H: Yet now you're holding you chin up. You're holding your chin up more than I have seen. You're compensating with your chin.

S: I'm compensating?

H: Be as natural as you can.

S: Yeah. I'm trying to get into it. What happens to me lately is that when I can contact the block, then I can start to release it and I begin to feel flowing in it. So, a lot of tension across the fronts of my legs. Tension around my eyes, at the moment. But especially the back.

H: O.K., Ron. Go ahead!

R: Well, the thing that strikes me immediately is the planted stance.

H: Right! I agree.

R: Planted in defiance. Real defiance. A slight twist to the right. The right's going back.

H: The knees are locked.

R: From the back he looks like he's saying, "You fuckers, I'll . . ." There's a real, "I won't!" built right into it. The legs are small. I see a lot of need for nourishment . . . you know . . . the legs have . . .

H: There's a collapse . . . there's a fight against collapse in there.

R: Yeah. There's a bigness here *(the upper body)* that's not reflected in the legs.

H: Right! In other words, his pelvis is really tiny compared to the size of his thorax.

R: There's a collapse right here *(the lumbar region)*. The thorax is jammed down into the pelvis.

H: It's like the pelvis and the lumbar spine are very short, compared to his whole thorax. And he's got . . . actually, the split there . . .

R: There's a bubble sense, the tension bubble in the whole trunk.

H: Right!

R: The shoulders are rotated inward. Actually, straight up and down is pretty good.

S: If I could lengthen this side of my body, by about three inches . . .

H: The front of your body?

S: No, this whole area here *(the chest)*. Then I'd be more in proportion.

R: The thing I get is that you can't move. From that position, you can't move.

H: He's planted. He's held. There's a rigidity in the legs.

R: Massiveness in the chest and back.

H: Not so much in the front of the chest. The back shows the massiveness to me. The back.

R: Yeah! The trapazoids are filled up here *(about shoulder-level)*.

H: Also, his buttocks are pulled up off the ground. They're tucked up, and they're pulled up between, and they're holding.

R: The hamstrings and the thigh have got to be very . . .

H: Well, you can see that the way that the legs are screwed into the pelvis is abnormal. That is, they're not on line, they're screwed around in there.

R: Oh yeah! They're out at an angle.

H: So, he's really holding on to that. His grounding is screwed in improperly.

S: That's the whole thing in holding things in, holding this whole area in *(the buttocks)*. 'Cause there's a fight with the mother and the fight has to do with eating and defecation, input and output. There's a fight up here *(the neck and shoulders)* and there's a fight down here *(the buttocks and pelvis)*.

R: There's some here too *(the lower belly)*.

H: He has some there, not as heavy as, let's say yours. He has some of that upper-lower block, too. Now the thing is his face, Sam's face.

R: Craggy. He looks like a prophet.

H: He looks like a prophet, he looks like . . .

R: Your face looks like you're saying, "You have done something bad." You're saying it to me. You know, like "You've done something terrible."

H: Look at the jaw now. Look at the holding in the jaw. And the lips.

R: They look sullen and defiant.

H: That looks sullen and a little defiant to me, too. And then, under it all . . .

S: There's a sweet guy.

H: There's a sweet guy under it all, but under it all there is also the . . . the . . . pleading. I don't know how to put it.

R: Oh yeah! The eyes.

H: There's a pleading under it all. And there's a demon, a striking-out aspect, a striking-at-you aspect . . .

R: Well, I don't think it comes out. I feel like he's not gonna hit me, but he's gonna tell me that I have done something terrible.

H: Right! It's there. It's really there, but I think there's also striking under there, too. There's the possibility. There's the hurt and a strike behind it.

R: I feel that the shoulders are too crushed in . . .

H: See, the lips are thin. The lower lip is thin. It's quite thin. There's the bitterness, in there. There's bitterness in the lips. Those are some of the things I see.

S: A mess.

H: We're all a mess.

R: Let's just go to the hospital and forget the book.

H: O.K., good.

R: The shoulders coming up and up.

H: Overburdened. We forgot to mention the rounding of Sam's shoulders. Very rounded. Overburdened. His back would make a very beautiful picture to show that. The rounding. Overburdened. Too much.

S: Lunchtime. Lunchtime, you guys.

We have a pretty clear picture here of Sam, a strong, forceful man. His feelings are being crushed down. He's crushing them down himself with his body structures. The tucked-under pelvis and rounded, massive back form a vise around the soft feelings which are there in the front of the body. The message in his face is one of defiance and, an accusation, "You hurt me." This can be seen as a response to feelings of defeat and suppression. Hector sees in Sam a demon and the possibility of striking out. Ron doesn't see that as clearly.

It has been contended by Freud, Reich, and many others that the average man goes about without much awareness of his deeper self. He is "unconscious." You can see this with our authors. Each one begins with his experience of himself and, even though he contacts some of himself, the others see more and differently. That's the way it is with most people.

7.

YOUR OWN
BODY PROFILE

This chapter gives you a chance to examine your own body to discover the attitudes and feelings related to its structural features.* To do this, we've illustrated four types of bodies.** These types represent four basic ways in which the body tends to differ from an ideal, normal structure. As it changes from normal, the body may either (1) slump as if tired (2) squeeze down as if ducking, (3) stiffen, or (4) become top or bottom heavy. For each type, there is a set of illustrations and a table which gives a structural and psychological description.

HOW TO USE THE ILLUSTRATIONS AND TABLES

By looking at yourself in a mirror or having a friend help you, you should find that one of the illustrations or a combination of several illustrations provides a reasonable approximation of your own body structure and posture. While most people are combinations, the types we have shown are pure forms, so you may have to look at different figures and parts of figures to get an accurate picture of yourself. For example, you may have a body that looks like a burdened type from the waist up and a rigid type from the waist down. If so, your psychological characteristics will also be a combination of these two types.

* For a fuller description of body types see Alexander Lowen's *Bioenergetics*, New York, Coward, McCann & Geoghegan, Inc. 1975.
** Illustrations provided by Lynda Braun.

One very useful way to understand the meaning of body structure and posture is to assume the various postures and see what you feel like in each one. We recommend you do this, using the illustrations as a guide to getting in the postures. The feelings that result should give you an idea of what people with these types of bodies experience.

The illustrations contain sequences of four figures each. The first figure in each sequence is approximately normal from a structural point of view. As any sequence progresses, the figures differ more and more from normal, becoming finally an almost pure example of one of the types.

Deposits of fat tend to mask the structural changes we are trying to illustrate. So, if you are heavy, you may have to imagine your own profile with less weight on it. There is a psychological meaning to the fat though. In general, fat can be seen as stagnant, unreleased energy. Similarly, a deficiency of weight can be seen as an inability to absorb or contain energy.

As you use the illustrations and tables, the following guidelines will be helpful:

(1) If you are normal looking, we don't have much to say about you, except perhaps that you are probably also psychologically normal.

(2) If you are one step from normal and look pretty much like Figure B in any one of the sequences, then your deviations from normal are mild ones and the psychological interpretations will apply only as tendencies or reactions under stress, when tired, or when irritated. The feelings and attitudes will be there, but not in a form which would override good sense, effective functioning, or your capacity to enjoy life.

(3) If you match Figure C in one of the sequences, you should have clear tendencies to experience yourself and your life in the ways described in the table. The situations in which you find yourself, the decisions that you make, and the general course of your life should all be colored and shaped by the feelings and attitudes outlined.

(4) If you look much like Figure D in any sequence, your character traits are deeply set. The description given for you

should match quite closely, though we expect some of it will seem alien or even distasteful to you. It might be helpful to see how someone you know well and trust feels about the description. When the structure is distorted to this extent, the defenses are rigidly maintained. You are a very definite way and you are that way almost all the time. The tendencies and attitudes that we would say are masking your true nature are the very ones you define yourself with and think of as "just you." For you, your distortions are yourself.

No body in the illustrations is beyond change. No matter what your age or type or combination of types, a great deal can always be done to free yourself of the tensions and beliefs that bind you to a distorted structure and an incomplete or unsatisfying emotional existence. We discuss this in detail in the next chapter.

TABLE 1. THE NEEDY TYPE

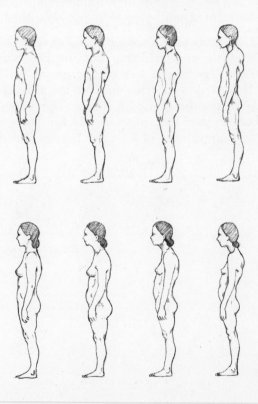

Main Structural Distortions: The body slumps downward and tends to become thin. The head is thrown forward, the chest sunken and the knees locked.

The Impression Given: The body looks tired and weak, in need of support.

Experience of a Person With This Type of Body: The needy type often finds it difficult to cope and may get discouraged or depressed easily. They will also have a desire for many friends, lots of social activity and much attention from others.

Characteristic Behavior Patterns: Most feelings are easily expressed with the possible exception of anger. This type seeks help and support from others and may appear childish and impulsive at times. There's a tendency to fall into dependent positions. The voice has a weak, sad quality.

Underlying Fears and Other Emotions: This person has a strong underlying fear of being abandoned, being left alone and helpless. There are deep feelings of emptiness and isolation and a strong resentment at not being held and nourished.

Structural Features and the Emotions They Represent: The thrust forward head shows a reaching out for nourishment. The rounded shoulders belie a lack of aggression, an inability to take what is needed or to strike out with the arms. The sunken chest holds deep sadness and loneliness. Tension in the abdomen blocks feelings of emptiness. The locked knees are used to hold the body up in spite of a lack of energy and strength.

TABLE 2. THE BURDENED TYPE

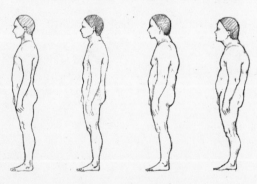

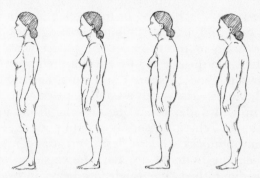

Main Structural Distortions: The body squeezes downward, shortening and thickening. The main tensions are in the flexor muscles, curving the body forward.

The Impression Given: The body looks as if it is under strain, as if it is carrying a heavy weight or stuck in an unavoidable and unpleasant situation.

Experience of a Person with This Type of Body: The burdened type often feels bogged down, getting nowhere in spite of great effort and struggle. They feel as if they are under pressure, so great at times that they could almost burst. They have a deep sense that they are suffering, a pervasive feeling of inferiority and a great desire to be close to others.

Characteristic Behavior Patterns: This type has difficulty expressing emotions and asserting themselves. They can be very stubborn and can persist in situations where most others would quit. There is a tendency to fall into submissive positions. The voice has a whining quality.

Underlying Fears and Other Emotions: This type has an underlying fear of being hopelessly stuck, trapped, or lost. There is a sense of not ever being able to be fully as alive and as "good" as others. Deep resentment and anger at being suppressed is also there.

Structural Features and the Emotions They Represent: The jaw may be thick and tense, showing effort and holding on. The neck will be short and thick, the head seeming to be pulled into the shoulder girdle. This reflects a fear of taking chances, a ducking in the face of an expected blow. Spite, resentment, and hopelessness are also blocked in the neck. The shoulders are rolled forward and are heavily built, holding back anger and taking the strain or the brunt of the punishment. There's a sense of defeat in the curving forward of the upper body. The front part of the body is shorter and tense, holding in feelings in general, but especially sadness and hopelessness. The pelvis is tucked up and under like a dog drawing its tail up between its legs. This is a characteristically fearful position, but, where the hips and buttocks are also fleshy, great anger is also held there. When the buttocks are flat, the capacity for physical pleasure is diminished. The thighs are heavy in front and are carrying the weight of the body. The hamstrings are very tense and are working with the abdominals to hold the pelvis in the tucked position.

TABLE 3. THE RIGID TYPE

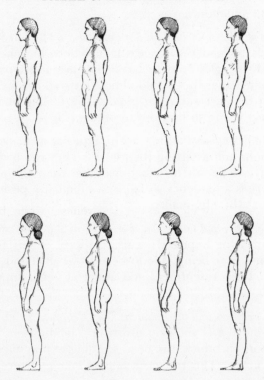

Main Structural Distortions: General, overall tension, more pronounced in the extensor muscles, curving the body backward. The neck and shoulders are held stiffly and the chest tends to stay inflated.

The Impression Given: The body looks stiff, at attention, or braced against a challenge of some sort. There's a sense of aggression and readiness to it.

Experience of a Person with This Type of Body: The rigid type will often experience a sense of frustration, a feeling of being opposed, blocked or challenged. There's a desire to succeed, to achieve and to be admired for one's accomplishments. This type has difficulty in relaxing, slowing down, and taking things in.

Characteristic Behavior Patterns: This type is generally active. They are also usually productive. They tend to be rational, logical, and serious, prone to thinking about rules, facts, technical matters, and details. The female of this type can be just the opposite, having a more global and romantically oriented mental life. Both male and female may be easily angered, aggressive, and will have trouble handling tender, soft feelings. The voice tends to be strong and deep.

Underlying Fears and Other Emotions: There is a strong fear of being suppressed or held back. A deep longing for love from the father is there along with anger at not being given recognition, love and support for one's own sake, without having to always fulfill other's expectations.

Structural Features and the Emotions They Represent: The strong jaw shows determination and aggressiveness, while holding back fear and impulses to cry. The stiffness in the neck and shoulders holds both anger and resentment and maintains that singleness of purpose that achievement seems to require. The often broad shoulders, thrown slightly back and up, show a readiness to assume the responsibilities of and a desire to be accepted as a full adult. The inflated chest holds sadness and longing, especially for tenderness, while giving an outward impression of pride, strength and independence. The pelvis is held in the cocked position, like the hammer of a gun pulled back and ready, but the person is unable to yield in this area to free-flowing, spontaneous movements. This is related to the tension in the lower back which holds the pelvis in that position. The buttocks are often well rounded and well proportioned, showing a capacity for physical pleasure. This capacity, however, creates a heightening of tension which, due to the inability to release the pelvis, leads to frustration in many cases. The back of the legs, the hamstrings, are tight and work in conjunction with the lower back to hold the pelvis in the drawn-back position.

TABLE 4. TOP OR BOTTOM HEAVY TYPES

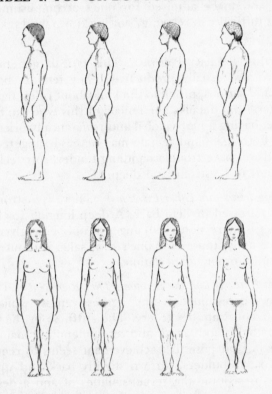

1. *The Top-Heavy Type (Usually Male).*

   Main Structural Distortions: The body expands above the waist and thins out below the waist. There is strong tension in the head, neck, pelvis, and legs.

The Impression Given: The body looks swollen up with pride or anger. It may appear frightening or even monster-like.

Experience of a Person with This Type of Body: The self-important type feels alienated from others and somehow outside the "normal" flow of life. There may be a strong desire for power over others or respect from others. A recklessness born of desperation and a disregard for painful realities will often permeate the life style of this type.

Characteristic Behavior Patterns: This type of person has a tendency to ignore the needs and feelings of others and himself. He is oriented towards manipulating others either forcefully, through fear, or by seductive maneuvers. He may have an exaggerated estimate of himself or his accomplishments. Many people classified by society as "deviants" are of this type. They tend to be opportunistic and impulsive.

Underlying Fears and Other Emotions: There is a deep fear of being overpowered by others, of being used. On a deep level, there is a struggle to survive as an independent self. Rage at being used and a longing for intimacy is also there.

Structural Features and the Emotions They Represent: The head is tense, partly because thinking, fantasizing and planning are overemphasized, partly because many situations are felt as struggles of will and this effort is held in the muscles of the head, and partly because rage and fear are blocked here and in the neck. Pride is expressed in the neck and in the blown-up chest. The arms and shoulders may be large, showing power and aggressiveness. The thinness and tightness in the pelvis, hips and legs show a diminished capacity for physical pleasure and realistic thinking as well as blocked rage at being used and fear of being overwhelmed.

2.   *The Bottom-Heavy Type (Female. Shown frontally for easier identification).*

Main Structural Distortions: The bottom half of the body is blown up out of proportion to the top. The hips, legs, and pelvis are large and the chest and shoulders are small, even for a normal body.

The Impression Given: The body has a doll-like or girlish quality above the waist and a full, sensual, womanly quality below.

These women are often very attractive to men and many artists —Titian and Renoir, for example—have preferred them as models.

Experience of a Person with This Type of Body: The experience of this type of person often centers on love relationships, home, children and sometimes status. Striving for acceptance, frustration and hurt feelings are mixed with a rich, warm, emotional and sexual life.

Characteristic Behavior Patterns: In contrast to the male of this type, the female is very concerned with feelings and hers are quite easily hurt. But, like the male, she also can be manipulative, using her girlishness at one time and her sensuality at another. She can also be very warm, receptive, and generous. She pursues situations that affirm her worth and womanhood and becomes upset easily, especially when she feels rejected. There is often a sexy tone to her voice and a suggestive quality to her movements.

Underlying Fears and Other Emotions: This person has a deep fear of being emotionally hurt, especially by being rejected by an important male love interest. There is anger at being rejected and a corresponding longing and struggle for acceptance.

Structural Features and the Emotions They Represent: The shoulders may be tight in an effort to block reaching out for love and acceptance which is associated with fear of rejection. This same fear results in a tightening of the chest, protecting, but at the same time, smothering the heart and narrowing the chest. The wide, ample pelvis is an attempt to compensate for the blocked heart feelings and to achieve real warmth and womanliness.

8.

LAST THOUGHTS

There are a few more things we'd like to just mention: other ways to look at the body; character types; and body-oriented appraoches to growth and change.

Many attempts have been made to systematize the connections between physical characteristics and personality. The shape of the head and face have been used, for example. A system which focuses on the distribution of tissue types is W. H. Sheldon's. The pure types are *endomorph*, *mesomorph*, and *ectomorph*, corresponding to viscera, muscle, and nervous system tissues. According to Sheldon, a person's character is dependent on the proportions of each type of tissue they have. An endomorph, for instance, has a high proportion of visceral tissue, and his or her character correspondingly centers around food and comfort.

Within the psychoanalytic tradition originated by Freud, *oral, anal, masochistic*, and *hysterical* are among the basic character types. Wilhelm Reich and Alexander Lowen have written a great deal relating these types to structural and dynamic features of the body. As in Sheldon's system, people are more or less combinations of types, though one trend will often predominate. With training and practice, a therapist will find such typing systems useful tools.

Therapy works to counter the mechanisms which stabilize suffering and retard growth. Those habits of movement, pos-

ture, and the control of feeling and expression that have led
to structural change in the body (holding blocks, compensations, and so forth) are, for us, the most important mechanisms. Body-oriented therapies work, in one way or another,
directly with these mechanisms.

In osteopathy and chiropractic, direct manipulations of the
body are the main tool. In each of these, the emotionally
based habits which lead to the physical problem are largely
overlooked. But there are approaches which use the body in
meaningful ways to relieve the suffering involved in what are
generally thought to be purely psychological problems.

Some therapies, like Gestalt therapy and psychodrama, focus on the reality of bodily statements, without much direct
physical intervention.

Of all the body-oriented approaches developed in the West,
six stand out as complete systems, with highly trained practitioners, a literature of their own, and a potential for strong
influence on the future. The first of these is Structural Integration (rolfing), developed by Ida Rolf. It is purely body-
centered. The next two, Reichian Therapy and Bioenergetics,
are gounded in psychoanalytic theory and were developed primarily by Wilhelm Reich, Alexander Lowen and their many
coworkers. They are so closely related that we can consider
them as one system. Patterning, the fourth, grew out of the
same theoretical perspective as rolfing and was, in fact, developed by Judith Aston in collaboration with Dr. Rolf. It utilizes
Rolf's concepts of line, symmetry and gravity. Systems five and
six are those originated by F. Matthias Alexander and Moshe
Feldenkrais.

In Reichian and bioenergetic work, the therapist, in addition to traditional analytic efforts, does a great deal of body
work. He will place the patient in stress postures to promote
energy flow, deeper breathing, and spontaneous movement.
He will use his fingers and fists to open blocked areas by
exerting pressure or by palpating or massaging certain areas.
And he will often encourage the patient to deepen the expression of his feelings by pounding, kicking, sobbing, or scream-

ing till these occur in a natural and spontaneous way. It is hoped that through this process deep changes in character can be effected, that an understanding and working through of negative feelings can be facilitated, that the tensions which segment the patient's body can be released, and a new integration accomplished.

The rolfers, for the most part, abstain from any psychotherapeutic intervention. Rolfing is done on the body. At least that is the theory. Since a lot of feelings and memories are released during rolfing, and since many rolfers are psychiatrists and clinical psychologists, some interchange often occurs. Generally, rolfing is totally body-oriented, a systematic attempt to realign the structure of the body and to integrate the myofascial system. The rolfing practitioner uses his or her fingers, knuckles, and elbows to stretch muscles that need lengthening, to separate muscle bundles that have become stuck together through improper use, and to stretch and move the fascial tissue that surrounds all muscle. It is this fascia that actually holds the body in its particular shape. Where the muscle is not used properly due to trauma and habit, the fascia become shorter and thicker and adhere to neighboring fascia. It is this process which makes it so difficult to change postural habits. So the rolfer works the fascial tissues in order to restore proper balance, coordination, and freedom of movement.

A minimum of ten one-hour rolfing sessions are given, with more if needed. These first ten hours follow a set routine during which layer after layer of fascia and muscle are stretched and realigned until the entire body has been covered. With some leeway for individual differences, the intricate process of restructuring the body proceeds step by step in an order that attempts to prevent regression into old patterns. In the process, breathing and energy level improve significantly. The emotional effects are often as dramatic as the changes in body structure.

In our experience, rolfing produces changes in a briefer time than any other system we are aware of.

In terms of teacher-student interaction, patterning is more

closely related to the Alexander and Feldenkrais methods. In all three of these systems, the practitioner offers guidance and positioning of the body, while the student learns new ways to move, breathe, relax, use, sense, and feel his or her body. The object is to break old habits, to increase body awareness and feeling, and to create new patterns of movement and stillness which are free of unnecessary tension and effort. The exercises employed involve only gentle movements, and the emphasis is on awareness and not on emotional expression.

The following material, reproduced here through the courtesy of the Guild for Structural Patterning, was contributed by Richard Wheeler, an advanced practitioner of Structural Patterning in North Hollywood.

Structural Patterning, developed by Judith Aston, evolved through the observation that individuals, whether or not they had undergone rolf processing, had more length in their bodies than they knew how to use. Sharing the premises of Dr. Ida Rolf's Structural Integration, Structural Patterning is a system for educating the individual toward a more efficient use of his body.

The experience of Structural Patterning begins with an analysis of the individual's familiar (and therefore preferred) patterns of movement. Walking, sitting, and standing are discussed and tied together into an integrated picture forming a base line against which the client may observe his progress and relate his understanding of the patterning work. Next the client is taught a series of movements that are designed for his structure to enhance his awareness of how to move each hinge more appropriately so as to relate to a more balanced pattern of movement. He takes this movement sequence home to work with. In later sessions the seuqence is expanded and sophisticated, as his understanding and awareness progress. Usually during the fifth or sixth session (normally there are between six and eight sessions) the work is related to everyday activities. This sequence may vary a great deal, depending upon the need and abilities of the client and later, if desired, the patterner may work with an individual's recreational or working activities.

Judith Aston has spent the last year working to make structural patterning available to people involved in all disciplines. This has involved creating advanced training classes in which trained practitioners are taught to work with individuals or groups involved with disciplines such as dance, massage, the martial arts, sports, and Yoga. The goal is to promote a more individual and efficient use of a person's structure within his chosen discipline. She has found that evoking a more efficient use of the body is of vital relevence to individuals who are involved with their structures on all levels. Children establish movement preferences very early in life and are shown alternative, more efficient patterns of movement through play, games created for their structures, or by gentle physical guidance. Pregnant women find great relief through learning to carry the baby closer to the more efficient vertical line of gravity, letting the pelvis provide support rather than stomach muscles or lower back. Musicians find they have more energy available for rehearsal and performance when they are shown easier ways to hold or play their instrument. Secretaries find their profession less stressful when they learn easier ways of sitting and of typing. Actors are able to add to their repertoire when they become aware of the basic pattern underlying their movement work. Psychotherapists of many diverse persuasions find it of great interest to relate body structure, movement preference, and posture to emotions or social situations. Another phase of the advanced training in patterning involves working with individuals with serious physical difficulties.

The overall goal of Rolf-Aston structural patterning is to give an individual understanding of and responsibility for his unique structure and movement preferences. It gives him the ability to balance his own body so as to minimize gravitational stress and maximize overall length and ease of movement.

The Feldenkrais method is described below by W. S. Dub Leigh and Betty Fuller, Feldenkrais teachers in San Francisco.

The body work created and developed by Dr. Moshe Feldenkrais consists of a one-to-one manipulation and a method of exercise. it is grounded in movement as the main means for individuals to educate themselves, improve their function as whole human beings, enhance and expand their self-image, and more fully realize their potential. Feldenkrais exercises consist of sequences of movement to be executed very slowly, gently, pleasurably, with sharply increased awareness of the developmental process Here/Now. The individual learns to correct inefficient and faulty movement by attending to his own feedback—physical and mental. By this means, one experiences the essential unity of the body/mind functions of the central nervous system.

Frequently an exercise is experienced on one side of the body by means of physical movement, and on the other side in the mind only—the student creating the movement sequence mentally, as if it were happening. Almost without exception, when the student actually moves, the side exercised only in the mind surpasses the other in achievement.

Feldenkrais exercises gradually eliminate from one's mode of action all parasitic or superfluous movements, and everything which blocks or interferes with movement. Our students experience their bodies as efficient machines operating with greatly reduced friction, increased tonus, and the optimum function of an organism correctly aligned in the field of gravity. The Feldenkrais method further results in increased sensitivity, heightened awareness, and expanded real sense of self. Unlike many exercise methods which promote unconsciousness by rapid, repeated strenuous effort, Feldenkrais exercises require full consciousness every moment, resulting in individuals who are fully able to experience their experiences.

The Alexander Technique is elaborated upon by Ilana Rubenfeld, a Gestalt and psychotherapist and certified teacher-trainer and member of the board of the American Center for the Alexander Technique.

The roots of the Alexander Technique grew directly out of F. M. Alexander's own process, which literally materialized before his eyes as he observed himself in a five-way mirror in the late 1800's. A Shakespearian actor plagued by a chronic loss of voice and doctor's failure to provide him with a permanent cure beyond "stop talking," Alexander began focusing his awareness on watching himself speak: he saw that each time he tried to speak, or even a thought of a word, he would throw his head back, his neck would come forward and there would be an accompanying shallow gasp for breath, resulting in a constant tension around his vocal chords. This discovery of a relationship between his head, neck and torso that he called "primary control" became the key to his work with his voice and to all subsequent work he did on himself. This opened the door to a process which became the pivotal point of his technique, central to which are the following:

1. Let the neck be free (which means that you see that you do not increase the muscle tension there in any act);

2. Let the head go forward and up . . . forward for the neck that is, not forward in space (which means that you see that you do not tense the neck muscles by putting the head back or down in any act);

3. Let the torso lengthen and widen out (which means that you see that you do not shorten and narrow the back by arching the spine).—F. M. Alexander, *Use of the Self*, Integral Press, 1955.

Alexander concluded that nothing we ever do is of local origin, but rather is interconnected with how we use the rest of our bodies.

We have learned to stand, to sit, to write and to walk in a way that most of us do over and over again, automatically, without awareness of "how" we are doing it. The first part of the Alexandrian work is becoming aware of our "hows." Awareness of "hows" requires time, self-observation and outside guidance. This includes observation of what we are doing and how we are doing it—which muscles do what, where we are holding. What we have learned to do, we can unlearn.

The next part of the Alexander experience is asking the student to do "nothing" (or what I call "finding your own *tao*.") This allows space where we can experience our bodies differently. Only after we allow ourselves to do this repeatedly from a neutral space can we eliminate habit patterns and replace them freely with alternate choices of movement and behavior.

The concept of the body as an organic whole is crucial to the Alexander Technique. We can go on correcting isolated parts of ourselves, but overall tension will continue to rebuild through our repeated, incorrect use of our bodies.

Alexander's discovery of 'primary control' led him to important insights into the relationship of our bodies with gravity. We are never still, even when quiet, for our center of gravity constantly changes. Once, in Maine, when I was in a boat on the way to an island, I saw two birds in the distance, standing. As I got closer, they seemed to be standing still, but as approached I saw them move from one foot to the other, shifting, shifting. It was beautiful to see this because they were in a state of constant movement that was so minute as to be almost invisible.

As long as we are moving from a central core, we can go into any position and make any movement freely, without being constrained by old habit patterns. The result is feeling lighter, balanced and more alert.

Part of the Alexander process in breaking habit patterns underscores the relationship between thinking and moving. Before we even begin to make a gesture, our decision to act has already set muscles in motion. It is at that precise moment— between the time we think and the time we move—that we can stop to establish a *conscious control* by saying: "No, I won't do it that way!" This take a willingness on our part to slow down that process of movement in order to focus our awareness, thereby stopping the old pattern—"breaking the chain" (Alexander called this "inhibition")—and allowing for different ways of moving, i.e., communicating. Alexander utilized the getting in and out of a chain—an ordinary, automatic phenomenon—to demonstrate inhibition and conscious control.

The Alexander Technique is a re-education essentially taught on a one-to-one basis by trained teachers. However, in the last few years, inspired by Moshe Feldenkrais, we have experimented successfully with extending Alexander's principles to groups. In our one-to-one work, the student develops a kinesthetic awareness through guided sensory experiences and attention to moment-to-moment processes, not final goals. Ideally, the kinesthetic experiences are both inner and outer. The student works from the inside, both mentally and physically, while the guiding hands and systematic instructions of the teacher direct awareness.

Not long ago I was working on a client's shoulders to broaden her chest and free her breathing. As I guided her shoulders, her arms lengthened. She became frightened and began to cry. When I asked her to fantasize about what would happen if her arms were longer, she answered that she then might be able to touch her genitals. Her voice and appearance had become much younger. When I asked her how old she was feeling at the moment, she said: "Two or three . . . in a crib." Later, she released this fear in a flood of screams and tears. She told me this was the first time since she had grown up that she had remembered her mother tying her hands to the crib with colored ribbons to prevent her from masturbating.

While the Alexander Technique has therapeutic value, it is not therapy. In the course of their training, Alexander teachers learn how to 'stay' with people who have released traumatic material. However, psychotherapy is frequently needed to enable the working through and reintegration of material that underlies distress.

On the deepest level, change *always* involves the body. A new attitude means new perceptions, new feelings, and new muscular patterns. Psychological and physiological change go hand in hand. Since our deepest traumas are imbedded in our guts and muscles, to free ourselves we must free our bodies. Yet we are more than just bodies. We are minds and spirits, feelings and imaginings. And though the body speaks, it must always be the whole person to whom we listen.

SOME BOOKS AND AUTHORS

We mention a few works here to afford the interested reader an opportunity to pursue our topic further.

Temperament, Character, Emotions, and the Body. The following focus on the psychological counterparts of such aspects of the body as shape, size, posture, movement, muscle tension, and energy flow. All include detailed theoretical discussions.

Baker, E., *Man in the Trap.* New York: The Macmillan Company, 1967.

Lowen, A., *The Language of the Body.* New York: Collier Books (The Macmillan Company), 1971.

The Betrayal of the Body. New York: The Macmillan Company, 1967; Collier Books (Macmillan Company), 1969.

Depression and the Body. New York: Coward, McCann & Geoghegan, 1972.

Love and Orgasm. New York: The Macmillan Company, 1965; (New American Library) Signet Books, 1967.

Bioenergetics. New York: Penguin Books, 1976.

Reich, W., *Character Analysis.* New York: Orgone Institute Press, 1949; Noonday Press (Farrar, Straus & Giroux) 1972.

Function of the Orgasm. New York, Noonday Press (Farrar, Straus & Giroux), 1973.

Sheldon, W., *Varieties of Temperament*. New York: Hafner, 1970.

        Atlas of Men. Darien, Ct.: Hafner, 1970.

The journal, *Energy and Character,* published in England by David Boadella.

Body-oriented Therapy. Books that deal with physiology, movement, body image, and muscle tension, and which provide systems for changing these, include:

Alexander, F.M., *The Resurrection of the Body.* New York: Delta (Dell Publishing Company), 1974.

Dychtwald, K., *Bodymind*. New York: Jove, 1978.

Feitis, R., *Ida Rolf Talks About Rolfing and Physical Reality*. New York: Harper & Row, 1979.

Feldenkrais, M., *Awareness Through Movement.* New York: Harper & Row, 1972

        Body and Mature Behavior. New York: International Universities Press, 1970.

        The Elusive Obvious. Cupertino, Ca.: META Publications, 1981.

Hanna, T., *The Body of Life*. New York: Alfred A. Knopf, 1980.

Pesso, A., *Experience in Action: a Psychomotor Psychology*. New York: New York University Press, 1973.

Rolf, I., *Structural Integration*. New York: Viking/Esalen, 1975.

        Rolfing: The Integration of Human Structures. New York: Barnes and Noble Books, 1978.

Schutz, W., *Here Comes Everybody*. New York: Harrow (Harper & Row), 1972.

Chinese Medicine. For thousands of years, the Chinese have been using a theory of life force flow for treatment and diagnosis. For detailed discussions we recommend:

The Yellow Emperor's Classic of Internal Medicine. Berkeley: University of California Press, 1972.

Austin, M., *Acupuncture Therapy*. ASI Publishers, Inc., 1972.

Muramoto, M., *Healing Ourselves*. New York: Avon Books, 1973.

Physics, Medicine, and Energy. Six authors who have done a great deal of thinking and experimenting in this area are:

Bateson, G., *Mind and Nature*. New York: Bantam Books (New Age Series), 1979.

Capra, F., *The Turning Point*. New York: Simon & Schuster, 1982.

Flanagan, G.P., *Pyramid Power*. California Pyramid Publisher, 1974.

Holbrook, B., *The Stone Monkey*. New York: William Morrow & Co., 1981.

Krippner S., Rubin D., eds., *The Kirlian Aura*. New York: Anchor (Doubleday) 1973.

Puharich, A., *Beyond Telepathy*. New York: Anchor (Doubleday), 1973.

ABOUT THE AUTHORS

RON KURTZ earned his degree in creative writing and physics at Bowling Green, Ohio, after which he wrote and taught in the fields of electronics and computers. He did his graduate work in psychology at the University of Indiana, where he was an instructor. He later taught at San Francisco State University. While holding a private practice in psychotherapy in Albany, New York, he gave workshops in Gestalt Therapy, body awareness, and encounter. He currently leads workshops in Body-centered Psychotherapy at the Hakomi Institute, which he started in Boulder, Colorado.

HECTOR PRESTERA, M.D., internist and cardiologist, is also a trained acupuncturist and rolfer. As a member of the teaching staff at Esalen Institute, Big Sur, he has conducted workshops utilizing Gestalt, body-energetics (Reichian-based therapy), encounter and Structural Integration (rolfing). His teachers have included Ida Rolf, Professor Jack Worsley, John Heider, John Lilly, Dick Price, Oscar Ichazo, Will Schutz, J. Moreno, Robert Monroe, and George Simon. He is currently developing a self-programming system that he has presented in several seminars under the heading of "Reprinting." He has a private practice in Monterey, California, where he continues to integrate Chinese traditional acupuncture methods with Western medicine.